C0-DBS-675

Policy Papers
in International Affairs

NUMBER 22

Nuclear Waste Disposal under the Seabed

ASSESSING THE POLICY ISSUES

Edward L. Miles
Kai N. Lee
Elaine M. Carlin

**Institute of
International Studies**

iiS

UNIVERSITY OF CALIFORNIA • BERKELEY

In sponsoring the Policy Papers in International Affairs series, the Institute of International Studies reasserts its commitment to a vigorous policy debate by providing a forum for innovative approaches to important policy issues. The views expressed in each paper are those of the author only, and publication in this series does not constitute endorsement by the Institute.

International Standard Book Number 0-87725-522-9

Library of Congress Catalog Card Number 85-80007

© 1985 by the Regents of the University of California

CONTENTS

LIST OF TABLES

ACKNOWLEDGMENTS

This study was funded by the U.S. Department of State, but the views, conclusions, representations, or descriptions are the responsibility of the authors alone and do not necessarily reflect the official position or policy of the United States government.

The authors would like to thank the following for their careful reading of the draft report and for providing cogent and helpful criticism: William T. Burke, Thomas A. Cotton, Kenneth R. Hinga, Todd R. La Porte, William Lounsberry, Gene I. Rochlin, Edith Brown Weiss, and Warren S. Wooster. None of these individuals should be held responsible for any shortcomings which may remain.

Chapter 1

INTRODUCTION

There is a growing literature assessing the scientific, technical, and policy implications of disposing of high-level radioactive nuclear waste (HLW) in general[1] and the option of sub-seabed disposal (SSD) in particular.[2] In this study we shall attempt to identify, evaluate, and strengthen understanding of the policy issues inherent in the SSD option, and we shall give most emphasis to the foreign policy issues. Technical information will be included where it informs or clarifies policy issues, but it is not our purpose to explore the technical issues in detail.

Specifically we shall attempt to answer five questions:

1. What are the problems of radioactive waste management? (Chapter 2.)
2. What are the technical considerations and uncertainties surrounding the SSD option? (Chapter 3.)
3. What organizational and operational problems may be posed by the creation of an SSD system? (Chapter 4.)
4. What domestic political problems for the United States may be encountered in the development of the SSD option? (Chapter 5.)
5. What are the likely international problems of creating and operating an SSD system? (Chapter 6.)

A permanent solution must be found for the very large and growing inventory of radioactive wastes which has been accumulating in the United States, Western Europe, and Japan since World War II. Furthermore, this solution must provide human beings protection from the high toxicity of the assorted radioactive isotopes—or radionuclides—some of which have half-lives measuring in the tens of thousands of years.* The SSD option is one in a series of alternatives

---

*The persistence of radioactive elements is indexed by half-life—the length

that must be evaluated in detail. There are three terrestrial options of high priority, but the SSD option is the only serious contender for the disposal of HLW in the marine environment.

The primary technical considerations and uncertainties surrounding the SSD option are the following: [3]

1. The effect of heat on the near-field sediments contained in disposal sites, and in particular the chemical response of these sediments;*
2. The potential effects of active, natural advection of pore water in and through near-field sediments;
3. The life of the canister;
4. The reliability and effectiveness of alternative emplacement systems for the canisters.
5. The risks involved in loading canisters on ships and transporting them to emplacement sites;
6. The risks involved in the emplacement of canisters;
7. The capability to retrieve canisters which have been emplaced;
8. The risks to the marine environment and to man after emplacement of the canisters.

Each of these considerations and uncertainties can be carefully assessed, if not resolved, by research which is currently under the auspices of the Nuclear Energy Agency (NEA) of the Organization for Economic Cooperation and Development (OECD).[†] However, one uncertainty will not be susceptible to such controlled assessment. It arises from the fact that for most countries pursuing a solution of the HLW problem, SSD will be at least a second-generation option to be considered after experience with terrestrial systems of disposal has been accumulated. The fate of SSD will therefore be much affected by the successes or limitations of terrestrial systems. As a result, it

of time it takes for half of the original radioactive atoms to decay. In ten half-lives a sample containing a radioactive isotope will lose 99.9 percent of its radioactivity from that isotope. At the concentrations found in military and industrial nuclear energy ten half-lives is often taken as a rule of thumb for judging when a given radioisotope has decayed to insignificant levels.

*Near-field sediments are about one meter or less away from the waste canister. Far-field sediments are more than one meter away.

†The NEA is a specialized unit of the OECD. Its task is to facilitate cooperation among member governments concerned with optimum development of

may not be possible to separate the SSD option from the contentious linking of two other issues—i.e., nuclear power growth and general solutions to the radioactive waste disposal problem. At the same time, if the terrestrial systems work, the cost of the SSD option will be a far more potent variable in the political decision to deploy the system or not. The relative risk assessments attached to terrestrial versus SSD systems will also be critical variables in that decision.

The analysis to follow indicates that the potential organizational and operational problems posed by the SSD option are indeed daunting, but they are not much different from those posed by the terrestrial options. The SSD option differs in only four respects from its terrestrial counterparts: (a) The need for a port facility; (b) The choice of transportation routes on land; (c) The need for one or more ships; and (d) The pre-disposal risks of operating and managing a sea transportation system. These factors do not pose problems which are different in kind or scale from those posed by terrestrial systems. The differences might have been greater if it was necessary to standardize SSD technology among a large number of countries, but on reflection there is no a priori reason why this should be required.

A detailed design of the management system must await further development and testing of the engineering design alternatives. However, we expect that whatever the specific components, long-linked or serially interdependent activities will lead to the introduction of vertically integrated networks of organizations. Moreover, since super-reliability is likely to be the major operating requirement of any waste disposal system and complexity will be high, there will be a need for building in considerable degrees of redundancy to protect against component failure. In any event, while it is possible to design an effective organization for the short term, considerable uncertainty surrounds any attempt to design organizations which must last long enough to monitor disposed waste, either on land or in the subseabed. Therefore organizational/operational requirements will probably not be a criterion in the choice between terrestrial versus SSD options; they are virtually the same.

The scope of major potential domestic political problems of an SSD option for the United States, however, is definitely broader than

nuclear energy. The range of NEA programs includes studies related to radioactive waste management.

that of terrestrial options. In one respect, the political issue will be identical—i.e., the management of nuclear wastes, involving problems of siting, radiological risks to humans, and probable effects on future generations (among others). In another respect, the SSD option will add an entirely new issue—i.e., the protection of the world oceans. Therefore the SSD option domestically will invite contenders to link it with a variety of nonrelated issues for their own tactical advantage.

Contenders will emerge out of a mix of types of actors, including the following:

1. The environmental and public-interest community;
2. The nuclear industry and commercial interest groups;
3. Federal, state, and local governments;
4. The scientific community; and
5. The public not already encompassed within any of the types identified above.

Each of these actors has and will have a variety of options, not all of which are clear as yet. However, we think that their positions will be determined by only a few major factors:

1. Whether SSD is to be regarded as a supplement or alternative to terrestrial systems;
2. The scientific and technical adequacy of the SSD option and, more important, people's perceptions thereof;
3. The extent to which anti-nuclear groups succeed in linking the question of nuclear power growth to the SSD option even after the two have been separated by the presumed prior emergence of terrestrial systems; and
4. The scope of coalition-building in which the various actors engage.

The most important potential political problems posed by the SSD option are international. These are also unique to the SSD option because most likely sites so far identified are in the international seabed area, beyond the limits of national jurisdiction. Therefore the SSD option poses difficult questions of feasibility, reliability, legitimacy, liability, and urgency.

The feasibility question—quite apart from a technical assessment of risks—largely concerns the political acceptability of SSD to the international community. Moreover, since the SSD option may be seen by a few countries as a first-generation technology, the international community will be called upon to make judgments about political acceptability before it is ready to do so.

One might ask whether it is necessary for the international community as a whole to participate in a decision of reliability, or whether it is enough that those who develop the technology be the ones who make that decision. Given the other potential international problems of the SSD option, we do not think that the small group of countries who have developed the technology will be sufficient to provide legitimacy to SSD because the likely sites are in global commons and the consequences of error are perceived to be potentially catastrophic to present and future generations.

The need for legitimacy will require negotiated agreement on who is to decide what issues. This, in turn, will pose questions of regime design and creation—in particular the sorting out of what needs to be done at regional and global levels. Moreover, decisions on who is to be held liable for what kinds of damage will have to be made before the fact. Finally, differing perceptions of urgency between some West European states and the rest of the world community will produce a certain amount of tension.

If the SSD option proves to be technically feasible at the international level, the major question will be whether conflict can be avoided over the design of the regime itself. In comparing the SSD issue with analogs in the international system, we found that conflict over regimes is highest where there are substantive links to resource allocation or the distribution of benefits and/or where principles of coastal state jurisdiction are involved. The SSD option can avoid such conflicts if it is insulated from other contentious issues that might conceivably be linked with it.

If the United States neither signs nor ratifies the third UN Convention on the Law of the Sea of 1982 (or the Law of the Sea Treaty) in the short to medium term, we think successful insulation will be unlikely and conflict over the SSD regime itself will be probable. It will be triggered by an asserted use of a disputed "common heritage" area. Conflict is likely to occur whether or not Western Europe and Japan sign and ratify the convention because those who sign and ratify will want the United States to participate and will seek to link a variety of other issues of concern to the United States to such participation. If the Soviet Union and its allies also sign and ratify the convention, the pressure on the United States will be even greater because the Soviets will want the United States to share the responsibility and costs of financing the International Seabed Authority.

The issue of regime conflict apart, the most likely and necessary link for the SSD option is with the issue of ocean dumping. We think it necessary that an amendment be made to the Convention on the Prevention of Marine Pollution by Dumping of Wastes and Other Matter of 1972 (London Convention)* to resolve the ambiguity concerning dumping. Whether or not links will be made to the emerging transnational anti-nuclear movement is an open question, but we think these can be avoided if sufficient attention is paid to the need for global legitimacy and if states do not pursue the option of dumping low-level radioactive waste in the ocean.

The major problems of negotiating a regime lie in site selection and regime design. It is important to realize that the first choice of site will have significant consequences for all others. The way it is handled could determine whether the regime itself becomes a highly contentious issue in the international system, even without links to UNCLOS III. Negotiating a regime design will require significant political mobilization given the potential number of actors involved, the nature of the issue, and the uncertainties surrounding the technology. This will not be a small undertaking.

At the regional level a joint facility already exists; what will be needed is a common policy governing technical operations. At the global level we think a common framework with limited but important tasks will be best; the tasks will be spelled out in the analysis to follow. However, we stress that no disposal of HLW in the areas beyond national jurisdiction be allowed unless it is permitted by a linked regional/global regime and executed in accordance with defined standards and procedures.

Even though for the United States SSD represents a second-generation (or later) technology, it makes sense that the United States continue to develop this option. Important issues of environmental protection and harm to man are at stake. Continued research and development of the SSD option would permit the United States to keep abreast of developments and allow it access on a mutual

---

*For details on the London Convention, see p. 52 below.

assistance basis to the research and development of others. Furthermore, development of the SSD option would help to preserve U.S. options relative to commercial reprocessing of spent fuel if the economics of such undertakings become more favorable in the future (see Chapter 3).

Chapter 2

RADIOACTIVE WASTE: HISTORICAL AND TECHNICAL DIMENSIONS

The potential of nuclear energy is increasingly unclear. Within the first twenty years of its discovery nuclear technology had become the linchpin of geopolitics, the frame of a bipolar world. As the clarity of that bipolar order gradually dwindled, the military utility and political value of nuclear weaponry became problematic—without undermining the thorny necessity of the weapons themselves. Beginning in 1954 with a U.S. declaration of a commitment to "atoms for peace," civilian nuclear energy enjoyed its own twenty-year period of hopeful prominence, only to founder in the midst of the high petroleum prices that should have guaranteed its success.

In discussing the historical and technical dimensions of the radioactive waste problem, we shall concentrate on the United States, the first nuclear weapons state and the largest user of nuclear power. Data on American inventories of radioactive waste are unusually complete, as is the documentation of policymaking. The passage in 1982 of the Nuclear Waste Policy Act (NWPA; P. L. 97-425, 42 U.S.C. 10101) created a comprehensive national program for management and disposal; its principal aim was the development of two geologic repositories for HLW near the end of the twentieth century. Before discussing the current waste disposal alternatives, we present an overview of radioactive wastes.

RADIOACTIVE WASTES: PRODUCTION AND CONTAINMENT

Radioactive wastes—waste materials containing unstable isotopes—are produced in two basic ways. *Fission products* are the fragments of fissile nuclei split in nuclear reactions. They have half-lives measured in decades,* and their radioactivity declines to innocuous levels in roughly seven hundred years.† *Activation products* are formed when neutrons are absorbed by atomic nuclei; dense concentrations of neutrons are found within nuclear reactors and when nuclear weapons are detonated. Many activation products have short half-lives, although one family of nuclei—the actinide elements—forms an important exception. These include uranium and the transuranic (heavier than uranium) elements; the transuranics are all formed as activation products. Materials contaminated with actinides require isolation from the biological environment for many thousands of years.

In principle the prudent management and disposal of radioactive wastes are not difficult. In practice serious misjudgments have created potential and actual problems of pollution by nuclear wastes that will be costly to address and perhaps impossible to solve in a satisfactory way. Such problems have seriously limited the acceptance of nuclear-power generation in the United States and other industrial nations.[1] Nonetheless, four decades into the atomic age, a sizable inventory of wastes exists, mostly of military origin. It is likely to increase substantially even if no additional nuclear power stations are built. If we are to avoid imposing an unwelcome burden on future generations, some means of permanent disposal must be chosen.

*Half-lives are defined in the footnote on pp. 1-2 above.

†Some fission products have very long half-lives. Perhaps the most important instance is iodine-129, half of which remains sixteen million years after it is created. The long lifetime of iodine-129 is further complicated because this element is very hard to immobilize chemically. Other elements do not pose the same sort of problem of persistent radioactivity and chemical mobility. A panel of the National Academy of Sciences (NAS) recently concluded that "all of the iodine-129 originally in the waste is discharged to the environment" eventually (Waste Isolation Systems Panel, *A Study of the Isolation System for Geologic Disposal of Radioactive Wastes* [Washington, D.C.: National Academy Press, 1983], p. 256. Cited hereafter as the NAS WISP Study.) The very long lifetime of iodine-129 means that the health risks it poses are very low but continue for long periods of time—millions of years. This sort of hazard is characteristically different from the risk of high levels of radiation released in the near term. The problem of containing and controlling iodine-129 releases accordingly poses an engineering challenge distinct from that of managing the rest of the fission products.

TYPES AND QUANTITIES OF RADIOACTIVE WASTES

Three types of radioactive waste require permanent isolation from the biological environment:

1. *Spent nuclear fuel.* This has been irradiated in a nuclear power station. It contains fission products, the remaining "unburned" uranium, and transuranic elements. Several of these components would be useful if removed, and thus spent fuel is not a waste in the ordinary sense. In the United States virtually all spent nuclear fuel is stored temporarily at the reactors where it was used.

2. *High-level waste (HLW).* Engineering procedures called chemical separation (in the military production of plutonium) or reprocessing (of commercial fuel) can be employed to remove useful material from uranium targets or fuel elements exposed to neutrons in reactors. What is left over is high-level waste, an intensely radioactive liquid that can be converted into solid form. In France high-level wastes have been solidified in glass, and research is underway in the United States to develop a similar process for wastes from the defense program.

3. *Transuranic (TRU) wastes.* The handling of spent fuel and HLW leads to the contamination of containers, protective clothing, and other equipment. Such waste materials are called transuranic or TRU wastes.* They have low concentrations of radioactive contaminants, but the radiation hazard will last for a long time because the TRU elements decay quite slowly. TRU wastes pose environmental risks in the long term—beyond five hundred years—comparable to spent fuel or HLW.

A fourth category, low-level waste (LLW), is defined as all other radioactive wastes. LLW is usually disposed of by burial in shallow trenches.

Spent fuel and HLW contain large quantities of fission products. These elements decay relatively rapidly, generating heat in the process; the heat emitted complicates the problem of handling these

---

*HLW also contains TRU elements. It is usually differentiated from TRU wastes in that it generates enough heat to require additional engineering measures in storage or handling.

wastes. Strontium-90 (Sr-90) accounts for most of the decay heat in the years immediately following removal of fuel from a power reactor.

There are approximately 8,000 metric tons of spent fuel in the United States, almost all stored at reactor sites. Over a 40-year lifetime, a 1,000-megawatt power reactor will generate 120 metric tons of spent fuel. The U.S. Office of Technology Assessment estimates that 150,000 metric tons of spent fuel will be generated by U.S. reactors in operation or under construction by 1990.[2] Table 1 provides estimates of the anticipated inventory of spent nuclear fuel from the 10 largest national users of nuclear power.

Table 1

LIFETIME COMMITMENTS OF SPENT FUEL BASED
UPON ESTIMATES OF NUCLEAR CAPACITY BY 1990

Country	Estimated Nuclear Capacity in 1990 (GWe)	Spent Fuel Commitment Implied (MTU)[a]
United States	50 - 150	50,000 - 150,000
France	45 - 60	45,000 - 60,000
USSR	30 - 40	30,000 - 40,000
Federal Republic of Germany (FRG)	30 - 40	30,000 - 40,000
Japan	25 - 35	25,000 - 30,000
United Kingdom	10 - 15	10,000 - 15,000
Canada	10 - 15	10,000 - 15,000
Sweden	10	10,000
German Democratic Republic (GDR)	5 - 10	5,000 - 10,000
Spain	5 - 10	5,000 - 10,000

Source: Capacity estimates from Irvin C. Bupp, "The Actual Growth and Probable Future of the Worldwide Nuclear Industry," *International Organization* 35, 1 (Winter 1981):61.

[a]Based upon a projection of 1,000 MTU per GWe for a 40-year plant lifetime. GWe: gigawatts of electric capacity; MTU: metric tons of uranium.

The United States inventory of HLW is almost entirely derived from military programs, having been produced in the manfacture of atomic weapons and in the fuel cycle for naval reactors. The current inventory is sizable: 293,000 cubic meters at the end of 1980 (the volume of a cube 66 meters on a side, or roughly a cubic city block). The radiological hazard of HLW is difficult to express in terms that are readily grasped. A committee of the NAS characterized the high-level defense wastes at Hanford, Washington, which account for two fifths of the HLW inventory, as follows: The single most hazardous constituent of high-level waste is Sr-90, an element that resembles calcium chemically and thus can be absorbed into bones. The Hanford Reservation, where the wastes are stored, is located in the valley of the Columbia River, whose annual flow of 100 billion cubic meters of water makes it the second largest river in the United States. To dilute the Sr-90 in the Hanford wastes to levels acceptable for human consumption would require the annual flow of the Columbia for a thousand years.[3]

HLWs are found in five forms—liquid, salt cake, sludge, encapsulated solid, and calcine (a loose, granular material). Wastes in the first three forms are stored in steel tanks at Hanford and the Savannah River plant in South Carolina; these account for virtually all the volume of HLW. At Hanford the intensely radioactive elements Sr-90 and cesium-137 (Cs-137) have been partitioned from some of the HLW and are held as solids in capsules. HLW derived from the reprocessing of naval reactor fuel is stored at the Idaho National Engineering Laboratory near Idaho Falls, Idaho, as liquid or as calcine. The activity levels of HLW vary significantly. Defense wastes in particular tend to be less radioactive than the liquids derived from reprocessing of commercial fuels.[4]

HLW at Hanford and Savannah River presents serious challenges. Some storage tanks at both sites are operating well beyond their design lifetimes. Twenty of the 152 tanks at Hanford have leaked, with a loss of 450,000 gallons of waste into the soil.[5] Thus far, fortunately, the soil characteristics and low rainfall at the site appear to have immobilized the spilled radionuclides.

Materials contaminated with even small amounts of TRU elements like plutonium can be dangerous for very long periods of time not only because the contaminants are long-lived, but also because they are highly toxic. In principle TRU wastes should be isolated from the environment for thousands of years, as is proposed for HLW. The hazards of TRU wastes were not reflected in public policy until 1970, when the U.S. Atomic Energy Commission (AEC) proposed to define them as materials containing more than 10 nanocuries of actinides per gram.* In 1981 the AEC's successor, the Department of Energy (DOE), proposed to change this definition to 100 nanocuries per gram for the defense TRU wastes, which it controls.

TRU wastes are stored at 13 locations in the United States, 8 at sites operated by the DOE. The DOE sites contain roughly 1,800 kilograms of TRU elements. There are major gaps in our knowledge of waste inventories, however. To begin with, the testing of nuclear weapons at the Nevada Test Site has contaminated an undetermined quantity of subsurface rock and soil. Moreover, there are uncertainties deriving from earlier waste management practices. Before 1970 materials that would now be classed as TRU waste were treated as LLW. Liquids bearing transuranics were routinely disposed of in cribs — covered trenches from which liquid waste percolated into the soil. At the AEC locations where TRU wastes were disposed of, substantial volumes of subsurface soil and rock were contaminated: up to 9 million cubic meters of soil are affected, mostly at Hanford.[6] Similar practices prevailed at commercially operated sites. Plutonium has apparently leaked into watercourses near a repository at Maxey Flats, Kentucky; the respository was closed in 1977.[7]

Most TRU wastes have been generated in military programs. In 1969 a bad fire at an AEC facility at Rocky Flats, Colorado, contaminated a large quantity of material with plutonium. This led the AEC to set the current definition of TRU wastes and to begin to package them in retrievable form. The largest part of the retrievably stored TRU wastes comes from the Rocky Flats fire and is now at Idaho Falls. TRU wastes are not now being emplaced at commercial facilities but are being held, like spent fuel, at the locations where they are produced.[8]

*The curie is the standard measure of the activity of a radioactive sample. It is defined as 37 billion nuclear disintegrations per second, the activity level of one gram of pure radium. The nanocurie is one billionth of a curie. Ten nanocuries per gram is the approximate limit of detectability of the actinide elements; lower concentrations of radioactivity are difficult to measure reliably.

CURRENT WASTE DISPOSAL ALTERNATIVES

The NWPA has outlined a program built around geologic repositories, mined cavities in carefully selected rock formations. These are the principal alternative for permanent disposal in the United States.[9] In addition, the NWPA has defined a program for development of monitored retrievable storage (MRS)— near-surface storage which could in principle be serviceable indefinitely.

In the geologic repositories program salt has been the most thoroughly studied medium, but studies are also being made in basalt, granite, and volcanic tuff. No final decision has been made on the medium of the first repository. Salt remains technically controversial for three reasons: it is highly soluble, and thus vulnerable if water enters the formation; salt itself does not immobilize radioactive atoms chemically, unlike some other candidate rocks; and salt is often found with oil and other valuable minerals, enhancing the likelihood of inadvertent intrusion into a repository.

Researchers opine that satisfactory models exist to describe the processes of radionuclide migration should a repository be breached.[10] However, neither the flow patterns of water carrying wastes through fractured rock nor the chemical interactions among wastes, water, and rock have yet been modeled.[11] More generally, it will probably be difficult to analyze the detailed behavior of the rock surrounding a repository, and the analysis will be made more difficult by the high heat loads expected from spent fuel or HLW. Thus painstaking site-specific field studies will be needed to confirm the suitability of any proposed repository.[12]

The basic design strategy for disposal will be the so-called multiple-barrier approach—i.e., the form in which the waste is put, packaging material for the waste, and the geologic medium will be chosen so as to reinforce one another.[13] The predictive models give no indication that a mined geologic disposal system cannot isolate radioactive wastes safely using this approach.[14] However, inadvertent breaching of a repository by future generations remains a significant issue.* Table 2 briefly surveys national research programs on waste forms and means of packaging wastes.

*Regulations issued by the NRC on the near-surface disposal of LLW are designed to protect public health in the event of an accidental breaching of the burial ground after it has been permanently closed.

Table 2

RADIOACTIVE WASTE ACTIVITIES BY COUNTRY AND/OR AGENCY

Country and/or Agency	Preparation of HLW for Disposal	Study of HLW Waste Forms	Spent Fuel Packaging	Preparation of LLW & Intermediate Waste for Storage or Disposal	Preparation of TRU Waste for Storage or Disposal
Australia	xx[a]	xx			
Belgium (including Eurochemic)	x	xx		xx	xx
Canada	x	x	xx	x	
CEC[b]		xx			x
France	x	xx		x	x
FRG	xx	xx		xx	xx
India	x			x	x
Italy	x			x	
Japan	x	x		x	x
Sweden	x	x	xx	x	
United Kingdom	x	xx	x	x	xx
USSR	x	x		x	

Source: U.S. DOE, Assistant Secretary for Nuclear Energy, Office of Nuclear Waste Management, *Nuclear Waste Management: Program Summary Document, FY 1981;* DOE/NE-0008 (1980), draft.

[a]Explanation of code: x = research and development activities underway; xx = U.S. interest in cooperative program already identified.

[b]Joint Research Centre of the Commission of the European Communities, Ispra, Italy.

15

Three variations on the multiple-barrier approach are worth mentioning. All attempt to compensate for the inherent uncertainties of any disposal method that relies upon geologic containment. First, an Australian group has designed an ultra-stable waste form called synthetic rock—a material that can hold radioactive wastes in a mineral matrix highly resistant to leaching.[15] A second concept has been developed in Sweden, where HLW and spent fuel are to be kept in interim underground storage for forty years. After four decades the heat emanating from these wastes declines by half, facilitating elaborate packaging in copper cylinders surrounded by bentonite clay, which adsorbs ions if there should be a leak. This waste package is highly resistant to radionuclide migration even if the geologic surroundings are less than perfect.[16] Third, J.D. Bredehoeft and T. Maini, geologists from the U.S. Geologic Survey, have proposed selecting sites that can provide multiple geologic barriers, taking into consideration both the formation where a waste is to be stored and its surrounding strata. In favorable cases the water flow patterns can be predicted for layered strata. Thus the means by which radioactivity can be transported back to the biosphere can be understood with confidence and a location selected where leakage, were it to occur, would be slow enough to pose no danger.[17]

Some locations already contaminated by radioactive wastes, such as Hanford and the Nevada Test Site, will probably never be returned to wholly unrestricted use. Accordingly these are "natural" candidates for repositories.[18] Indeed exploratory work on basalt is being done at Hanford, and granite and volcanic tuff are being studied at Nevada. Already contaminated sites pose the most serious long-term environmental challenges.

IMPLEMENTING GEOLOGIC WASTE DISPOSAL

The emphasis in the NWPA on the development of mined geologic repositories commits the United States to a national solution to the problems of radioactive waste. In the NWPA context, SSD is an alternative path of technical development which is likely to play a significant part only if the principal path is blocked.*

*SSD is not mentioned directly in the act. There are two indirect references: Section 5 explicitly adopts the strictures on dumping of radioactive waste

However, the uncertainties of terrestrial disposal are sufficiently great that the understudy's role remains vital to the goal of finding a feasible and socially acceptable means of disposing of radioactive waste. Here we present a brief summary of the uncertainties connected to geologic disposal.

First, geologic repositories may prove unworkable from a technical standpoint, although the probability of technical failure appears to be low. The NAS has concluded that "The development of repositories in candidate geologies . . . is feasible in terms of present construction and mining technology."[19] These words reiterate the conventional scientific wisdom that nothing has so far raised serious concerns about technical feasibility. This technical consensus is an important basis for optimism, though it is not a guarantee. In particular, the site-specific investigation of potential repositories has yet to be carried out, and significant engineering problems—such as the plugging and sealing of access tunnels—have yet to be solved.

A second, less predictable uncertainty is the social acceptability of a site and the design developed for it. The NWPA includes two procedural elements that suggest the character of the problems: a tight time schedule and the right to reject a repository site. Congress adopted a tight time schedule for repository development. The first repository is to be in operation by the early 1990s.[20] Even if the first repository has not opened, the federal government is directed to accept spent nuclear fuel by 1990.[21] The schedule is a political assertion that a solution to the radioactive waste problem exists. An overall mission plan is required to assist the DOE in identifying potential schedule disruptions.[22] Although the technical problems of repository development may be soluble—as the scientific consensus holds—solving them on a tight time schedule may prove impossible. This would in turn impair the credibility of the DOE—a credibility that will be needed to buttress evidence that the sites chosen will pose little or no risk to public health or the natural environment.

While Congress has sought to assure the nation that a safe repository can be located and designed, it has through the NWPA also preserved the right of states and Indian tribes to disagree with

contained in the Marine Protection, Research, and Sanctuaries Act of 1972, and Section 222 calls for "research on alternatives" to geologic repositories.

the DOE over specific sites. A potential host state or tribe can veto DOE's nomination of a site.[23] This veto can be overridden, but it requires majority support in both houses of Congress.[24] The veto provision of the NWPA acknowledges the legitimacy of "not in my backyard"—the desire to avoid a concentration of risk when the benefits have been widespread. How the veto would work is not clear; the possibility of deadlock cannot be ruled out.

As we have noted, the NWPA represents a major institutional step. Yet the technical and social uncertainties that accompany the development of a radioactive waste industry are significant. There is no guarantee of technical feasibility, and there is at best a clouded outlook for credible implementation of the act.

Outside the United States, Western nations using nuclear power face similar challenges, and generally their geography and geology permit even less reliance on geologic isolation. Moreover, public opposition to nuclear energy in Western Europe and Japan threatens to block plans to proceed with radioactive waste disposal.[25]

In sum, although geologic isolation within the continental land mass of the United States is the leading alternative for radioactive waste disposal, questions remain.

Chapter 3

SUB-SEABED DISPOSAL: TECHNICAL CONSIDERATIONS
AND UNCERTAINTIES

The sub-seabed disposal of radioactive wastes combines several promising features: the science and engineering are relatively unexplored and thus offer tempting challenges to top researchers, and the distant, international location of a sub-seabed repository may provide a means to short-cut some of the controversy that has dogged radioactive waste policy. These potential strengths have been conspicuous by their absence in the research and development programs for terrestrial disposal. However, a clearer sense of whether the potential of SSD can be realized depends on an understanding of the technological and social resources needed to bring SSD to operational fruition. The context in which SSD emerges will affect the availability of these resources. In this chapter we shall examine technical considerations and uncertainties of SSD, comparing them to mined geologic repositories where appropriate.

BACKGROUND TO SSD: MARINE DUMPING

From the technical and (perhaps) legal standpoints, SSD is not dumping but emplacement beneath the sea floor. Nonetheless, the policy issues and political dynamics of the SSD program are affected significantly by past and present practices in radioactive waste disposal at sea. We briefly summarize these practices here.

The dominant human contributions to marine radioactivity have been military: sunken nuclear-powered vessels and the testing of nuclear weapons. At least three nuclear submarines have been lost at sea, two of U.S. origin and one Soviet. Although the nuclear fuel inventories aboard these boats have not been publicly disclosed, they are estimated to have ranged from 100 to 1,000 million curies—

roughly the size of the radioactive burden in a large power reactor.[1] Measurements by the U.S. Navy at the sites of the two American boats indicate no radioactivity release more than a decade after the sinkings.[2] The atmospheric testing of nuclear weapons, which declined after the Limited Test Ban Treaty of 1963, has contributed several tens of millions of curies of fission products to the marine environment.[3] These uncontrolled depositions are larger by an order of magnitude than routine releases.

The largest routine releases to the marine environment originate from nuclear fuel reprocessing facilities in the United Kingdom and France, which dump several hundred thousand curies a year into coastal waters.[4] By contrast, the total activity in the LLW dumped by all nations has amounted to less than 500,000 curies to date.

LLW dumping at sea is still carried out by several European nations. In the United States dumping permits were not issued after 1961, although dumping continued until 1970. Before 1970 approximately 86,000 containers—mostly 55-gallon steel drums— were disposed of at four locations on the U.S. continental shelf. U.S. policy is now framed in accordance with the Marine Protection, Research and Sanctuaries Act (MPRSA) of 1972. The dumping of HLW is prohibited, and LLW must be contained so that it decays before it can be released into the water column. In addition, the U.S. NRC requires a showing that ocean disposal is the method least harmful to human beings and the environment before a permit can be issued.[5]

At the end of 1982 the U.S. Navy issued a draft environmental impact statement discussing alternative methods for disposing of the more than 100 nuclear submarines to be taken out of service over the next 30 years.[6] One suggested alternative was the disposal of submarine hulls at sea, an action which would mark the resumption of U.S. dumping of radioactive wastes.

Dumping of LLW at sea has elicited strong opposition. The scientific basis of concern has been summarized by W. Jackson Davis.[7] Political action has been taken by the Pacific island nations,[8] and more recently by a coalition of environmental groups challenging the U.S. Navy proposal to scuttle submarine hulls at sea.[9]

SUB-SEABED TECHNOLOGY

SSD was initially proposed and developed by ocean scientists in the United States, but since 1976 an international Seabed Working Group (SWG) has met under the aegis of the NEA. SWG now has representatives from nine OECD nations and the Commission of the European Communities. The nations represented in SWG account for more than three quarters of the worldwide installed nuclear capacity.[10] Much of the scientific and engineering work of the SWG is funded by the U.S. DOE in the Seabed Disposal Program (SDP) coordinated by the Sandia National Laboratories.* The cost in fiscal 1983 was $6 million.[11] The United States is a party in a number of bilateral agreements bearing on radioactive waste; a list is presented in Table 3.

ORIGINS OF THE SSD IDEA

The concept of SSD emerged from a chance encounter among a chemical engineer, William P. Bishop, a marine geologist, Charles D. Hollister, and a scientific administrator, Philip Smith, in Washington in 1973.[12] Hollister and Smith had worked together on an expedition to Antarctica funded by the National Science Foundation, where Smith was a program officer. At dinner at Smith's home Bishop, who then worked at the Sandia National Laboratories on radioactive waste, described the frustrations of the AEC's search for geologic repositories on land for waste disposal. Hollister asked whether there had been research into the other two thirds of the planet's geology—the oceans. He knew of the work of his colleague, Vaughn T. Bowen, at the Woods Hole Oceanographic Institution which had shown that radioactive material from weapons testing was

---

*Research on SSD began in 1973. The SDP's major goal is to evaluate the SSD option and, if feasible, to develop an SSD operation. Annual funding has ranged from $.3 million in 1975 to $7.5 million in 1980. To date the program has enlisted more than 100 scientific and technical investigators from various universities and research centers. For more information on the SDP, see Sandia National Laboratories, Seabed Programs Division, *Subseabed Disposal Program Annual Report, January to December 1980*, vol. 1, *Summary;* SAND 81-1059 (Albuquerque, NM, 1982). Referred to hereafter as Sandia, *Annual Report... 1980.*

Table 3

U.S. BILATERAL AGREEMENTS ON RADIOACTIVE WASTE

Partner Country	Date Agreement Concluded	Principal Subject(s)
Australia	Agreement under discussion	HLW immobilization; Mine/mill tailings; Nuclide migration
Belgium	Agreement pending	Radioactive waste management
Canada	8 September 1976 (expired September 1980)	Radioactive waste management and systems analysis of heavy water reactors
European Community	Agreement under negotiation	Waste management
FRG	20 December 1974 (expired December 1979; extension pending)	Radioactive waste management
Japan	31 January 1979	Fast breeder reactors[a]
Sweden	1 July 1977	Radioactive waste storage; Stripa mine tests
United Kingdom	20 September 1976	Fast breeder reactors[a]

Source: U.S. DOE, Nuclear Waste Management.

[a]The agreements include an annex covering cooperation in nuclear waste management.

firmly trapped in deep-ocean clays. Furthermore, the plate tectonics model of global geological structure indicated that the centers of the plates should be unusually stable geologically.* Bowen's work pointed to the deep seabed as a scientifically promising location for isolating radioactive wastes.

Bishop found the idea intriguing. He invited Hollister to Sandia to give a workshop, and the SSD concept—together with an outline of the research needed to evaluate its feasibility—was roughed out.[13]†
D. Richard Anderson, a chemist and oceanographer on the Sandia staff, was invited to work with Bishop and Hollister; Anderson currently directs the SDP.

From the outset the SSD concept was modeled on academic science. Hollister felt the concept could be proved to have merit only if it survived a rigorous scientific peer review. Detailed exploration of the centers of the ocean-floor plates had not been done; there had seemed little of interest there to the geologist— precisely because the environment was so stable, the cost of working at great depths was great, and there were few economic incentives to explore. Yet research in the mid-plate regions might be rewarding academically because it called for original scientific work in the field of paleo-oceanography. By choosing an academic rather than engineering or bureaucratic model, Bishop and Hollister created an ambience in the SSD program radically different from that prevailing elsewhere in the U.S. radioactive waste effort. The program's principal strengths have been the disciplined skepticism of anonymous peer review, the inventiveness of academic entrepreneurs, and a natural inclination toward scientific collaboration across national boundaries.

---

*The plate tectonics model, which gained general acceptance among geologists during the 1960s and 1970s, envisions the surface of the globe as a set of rigid plates, each the size of a continent or ocean basin. These plates move through geologic time, carrying continental land masses with them. At the edges, where one plate collides with its neighbor, the theory predicts earthquakes, volcanoes, and other signs of violent geologic activity. The so-called "ring of fire," the volcanoes and earthquake zones surrounding the Pacific Ocean, is explained by the plate tectonics model. The centers of plates, far from the active edges, are expected to be very quiet geologically. This is a key part of the SSD idea, as explained below.

†It is a fortuitous coincidence that at that time Sandia, like other national laboratories, was feeling pressure to diversify.

TECHNICAL FEASIBILITY OF SSD

SSD would use the clays of the deep seabed as a geological medium for isolating HLW. The wastes, packed in metal canisters, would be buried in the sediments 20-50 meters below the sea floor. Studies completed thus far support the initial conjecture that the sediments provide a degree of isolation equivalent to that obtained in the terrestrial media under study. The U.S. SDP is working through a sequence of experimental and analytical studies to assess the technical feasibility of SSD by the late 1980s.

Far from the edges of the continental plates, the ocean floor is (as we have noted) seismically stable. Among the mid-plate areas there are extensive regions that are also oceanographically stable — far from the great ocean current systems, or gyres, that flow near the continental margins. Far from the coasts, with their influx of nutrients, the ocean is also a desert: both plant and animal life are sparse. This slows the rate of sedimentation on the ocean floor, but it also lessens the potential biological impact of inserting highly radioactive wastes into the seabed. As a consequence of uninterrupted deposition for many millions of years, the sediments are thick, and it is possible to make predictions for the long-run behavior of a subsea repository with greater confidence than for a terrestrial disposal facility. (All terrestrial sites are exposed to erosion, complicating predictions of long-term isolation).

SEAWATER AND THERMAL PROPERTIES

The seabed itself consists of a layer of clay, some 50 meters thick, which gradually hardens into sedimentary rock. The clay is saturated with seawater. The properties of the seabed reflect its unique combination of high pressure — the weight of the ocean — and sediment chemistry. These properties suggest that HLW buried 20-50 meters within the sediments would be securely isolated.

A primary strength of the seabed clays is their chemistry. The clays over most of the deep seabed are strongly reducing — i.e., they immobilize, or sorb, most of the ions found in radioactive wastes. This is the critical element of geologic isolation by the sediments.*

---

*Several of the terrestrial media being studied also have strong sorptive capability. As we stated in Chapter 2, salt is a notable exception.

Even though waste containers would leak some decades after emplacement, the clay would slow the migration of the radionuclides into the water above.

Slowing the movement of radioactive materials chemically will do little good, however, if they migrate toward the water by other means. Here the extremely high pressure of the deep sea plays an important role. At high pressures water does not boil; indeed heat moves through water or water-saturated clays by diffusion rather than convection. Diffusion is a process that can be modeled precisely, while convection — such as the rapid movement of water in a boiling pot — is difficult or impossible to describe in mathematical detail. Calculations of the behavior of a waste canister in the deep sea clays indicate that the wastes will stay put once they are buried in the seabed. Field experiments during the late 1980s will make measurements in the deep sea environment to check the computer models used in these calculations.

Computer-model estimates of the behavior of the wastes form the basis for long-term predictions of the stability of the whole subsea repository. It is impossible to demonstrate geologic isolation on a scale of thousands of years; thus model calculations play a critical role in the assessment of waste disposal alternatives. In this assessment the unusual design of a subsea repository suggests some important advantages over a terrestrial repository.

For one thing, SSD will limit the thermal effects of wastes. In a terrestrial repository once the wastes are in place, they release heat into the rock mass in which they are embedded, gradually raising the temperature of the rock. As the radioactivity decays in the wastes and the heat diffuses through the rock mass, the temperature of the rock subsides. The maximum temperature to which rock will be exposed is now planned to be about 250 degrees Celsius.[14] While the corrosive properties of seawater limit the acceptable waste forms (see below), the intimate contact between the wastes and the surrounding water will reduce the thermal effects resulting from heat generated in the wastes. The maximum temperature now planned is 200 degrees Celsius.

For another thing, because of the subsea thermal properties, it will be only a matter of years before the heat from a canister reaches the environment — the water column in this case; in a dry repository the corresponding process will take centuries. As a result, it may be possible to do detailed studies of wastes in a seabed environment during the period of greatest scientific uncertainty, when heat is available to drive unexpected chemical and physical interactions in the near-surface sediments—the so-called far-field geology. Far-field stability is the fundamental element of permanent isolation of wastes, so the ability to assess far-field effects more clearly in SSD is a significant advantage.

Finally, mid-plate/gyre regions are large in area, making it very unlikely that repositories would materially reduce the resource exploitation potential of the deep seabed. The probability of inadvertent intrusion is also low because repository sites can be selected from areas containing few valuable mineral resources. Preliminary site surveys are being conducted in the U.S. SDP.

It should be emphasized that SSD is little more than a promise at present. Because of its remoteness and lack of biological or economic value, the deep seabed has been explored only in recent years, and these explorations have been motivated in large part by the growing interest in waste isolation. Most of the scientific understanding today is based upon simulation models rather than empirical measurements.[15] Thus much remains to be learned before the practicability of SSD can be clearly assessed. For example, the chemical interactions between radioactive wastes and subsea sediments could prove to be surprising. The high pressure at depth, combined with the relatively high temperatures planned,[16] will create a physical environment that is hard to duplicate under laboratory conditions.[17] In situ studies are likely to be needed. An experiment to investigate heat transfer in the seabed clay is now in progress in the U.S. SDP.[18] Additional scientific uncertainties include the speed at which deep-ocean waters move, the composition and behavior of biological communities in the deep ocean,[19] and possible changes in the chemistry of seawater itself under the high pressure and temperature conditions created by the emplacement of HLW in the sub-seabed.

TRANSPORTATION TO AND EMPLACEMENT IN THE SUB-SEABED

Getting radioactive wastes into a sub-seabed repository poses novel engineering challenges. Wastes, packaged into slender canisters 3 meters long and 30 centimeters in diameter, could be put into sub-seabed clays at depths 30-200 meters below the sea floor using a variety of techniques. These range from free fall—that is, dropping wastes from a ship at the surface—to lowering them into a drilled hole. Emplacing waste containers in a controlled fashion from a platform floating 5,000 meters above the repository is not simple, although the technologies of off-shore drilling and exploration for oil and gas suggest that there are no fundamental problems to be solved. Studies are being made of the physical and biological pathways which could expose human populations to radioactive materials in case of an accident.[20] Developing an assured capability to retrieve a canister could be more difficult.

The system for transport and emplacement may face greater difficulties at the landward end than at the repository. Facilities for marshaling substantial quantities of radioactive wastes are unlikely to be welcome tenants in urban ports, and the procedures for safe embarkation through harbors with heavy marine traffic have yet to be designed. Military bases may be suitable sites for housing the port facilities since they are already owned by the national government, have security arrangements against sabotage, and are served by land transports which are able to ship heavy shielded casks.

Transport vessels themselves appear to offer no basic problems: the land vehicles and containers already developed can be used, and a design for a ship capable of carrying radioactive waste canisters and associated emplacement equipment would be straightforward. It is worth pointing out, however, that while the judgment that transportation and emplacement would not pose fundamental problems is widely shared among scientists and government policymakers, much of the work in ocean engineering—including port facility planning and ship design—has been deferred until issues of scientific and environmental feasibility have been settled.

The SSD system must work smoothly at a scale that seems quite large by today's standards. The projected rate of generation of spent nuclear fuel in the year 2000 would require the final disposal of one waste canister per hour, 24 hours a day, 365 days a year.[21] It is a commonplace of technological history that scales of operation that seem outlandish in one era are thought mundane in another; yet growth to full scale is typically achieved by trial and error. The magnitude of errors to expect is not clear.[22]

WASTE FORMS AND CONTAINERS

If it could be assumed that fuels from commercial nuclear power plants would be reprocessed before emplacement, then the form of HLW could be tailored to the sub-seabed environment. Yet the evolving economics of nuclear power indicates that reprocessing to extract plutonium and uranium from wastes is of doubtful feasibility. Without reprocessing the waste form would most likely be spent nuclear fuel. It should be possible to pack spent fuel rods into a canister suitable for initial emplacement in the deep seabed, but the chemical interactions between wastes and seawater must then be taken into consideration.

Seawater, including that trapped in seabed clays, is highly corrosive. It is unlikely that any metallic container would survive beyond several centuries. After the container is breached, the waste form and the surrounding clay comprise the principal barrier against entry into the ocean waters. The water column itself constitutes a secondary barrier that must be crossed before the radioactive materials enter the biological environment fully, but U.S. planners assume that dilution in the oceans cannot provide satisfactory isolation.[23] Model calculations indicate that the migration processes now projected will proceed so slowly that no leakage into the water will occur until the radioactivity has decayed.[24] In situ studies could of course reveal migration processes different from those used in the models.

Given these factors, the technological feasibility of SSD may be linked to decisions on waste forms and reprocessing. Moreover, the acceptability of any technology that *plans* on breaching the primary waste barrier may be problematic within the regulatory frameworks of some nations. Thus the institutional context in which disposal in international territory is licensed may be a major consideration. If reprocessing could be assumed, SSD might be more readily accepted. It is hard to adopt such an assumption now, however, in light of the unfavorable economic prospects of reprocessing. These prospects seem likely to shift only if there is widespread adoption of nuclear power and thus a much larger nuclear economy than is considered likely today.

SSD AS A SECOND-GENERATION OPTION

As we have indicated, SSD is not likely to be the first technology used for permanent disposal of radioactive waste.* The terrestrial geologic option has proceeded to engineering development in several nations already (see Tables 4 and 5). Moreover, the uncertainties associated with the risks from radioactive wastes favor techniques that allow for retrieval of wastes. Retrieval appears to be a simpler matter for a terrestrial repository. In the United States the NWPA assumes that disposal on land will be developed first. Thus it seems reasonable to think of SSD as a technology that will be developed and implemented in a setting affected by the successes and limitations of terrestrial disposal. This factor shapes the technological context of SSD, much as the international character of the subseabed shapes the political context.

The most important consequence of the second-generation status of SSD is that the linkage between nuclear power and radioactive waste disposal (in its present form) may be largely discounted in system planning and research. Much of the controversy surrounding radioactive waste policy in the United States is based on a perception—shared by the general public and pro- and anti-nuclear activists—that radioactive waste is the most serious unsolved problem in the development of commercial nuclear power. This perception has led to intense pressure by pro-nuclear interests to accelerate development of a repository to demonstrate the technical feasibility of disposal. Nuclear opponents have raised scientific and jurisdictional questions, some of them meant to impede progress toward final disposal of any wastes. The turbulence generated by this political battle has contributed significantly to the difficulties of developing techniques for terrestrial disposal. At the same time of course the rising salience of the waste controversy has been one of the major motivations for supporting the development of alternative technological options, including SSD.

---

*There are nuclear nations with little or no land area suitable for radioactive waste disposal; these include island nations, like the United Kingdom and Japan, and small nations with little suitable geology, like the Netherlands. These countries could use SSD as their national means for disposal, but this seems unlikely to precede disposal in terrestrial repositories in other nations, including the United States.

Table 4

MAJOR WASTE HANDLING AND ISOLATION ACTIVITIES BY COUNTRY OR AGENCY

Country or Agency	Transportation	Interim Storage	Shallow Land Burial	Geologic Isolation	Seabed Isolation	Airborne Waste Immobilization	Safety/Risk Analysis
Austria				x			x
Belgium	x[a]	x		xx		x[b]	x
Canada		xx		xx		x	x
CEC[c]							xx
Denmark				x			
FRG	xx	x		xx		xx	xx
France	xx	x	x	xx		x	xx
India				x	xx		x
Italy			x				
Japan	xx			x	xx	x	x
Netherlands				x			x
Spain		x		x			
Sweden		x		x			x
United Kingdom	x	x		x	xx	xx	xx
USSR	x	x	x	x		xx	xx

Source: U.S. DOE, Nuclear Waste Management

[a]Explanation of code: x = research and development activities underway; xx = U.S. interest in cooperative program already identified.

[b]Eurochemic/NEA Program—Mol, Belgium.

[c]Joint Research Centre of CEC—Ispra, Italy.

It is important to stress that the political value of the technical concept of SSD is likely not to depend upon whether SSD helps or hinders public acceptance of nuclear power;* that question should be settled by the success or failure of a terrestrial disposal program.† By the same token, a dramatic shift in the economic and political appraisals of nuclear energy could rekindle efforts to undertake reprocessing. As described above, such a change could moot the question of whether spent nuclear fuel is acceptable as a waste form, changing the outlook for SSD.

A second-generation technology like SSD therefore faces social expectations somewhat different from those that face terrestrial disposal.** If terrestrial isolation works satisfactorily, the comparison of the economic costs of SSD and terrestrial disposal may become far more important than is now the case. In addition, the levels of assurance needed with respect to accidents, routine exposure, and institutional reliability are all likely to be higher for SSD than for terrestrial disposal. Holding SSD to a higher standard will not be easy, however. Economic and regulatory comparisons would be complicated by the fact that a terrestrial system would already be in operation—though perhaps not in the United States—but SSD would not yet be. The probability that the acceptability of SSD would be based upon criteria more stringent than those applied to the terrestrial option produces pressure for more innovation than is the case with mined geologic repositories. The fact that SSD has already furnished scientific challenges attractive to leading oceanographers and engineers is a promising sign.

*As we discuss below, such a basic change in the political environment makes it difficult to estimate which issues will arise in the development and implementation of SSD.

†The failure of a terrestrial repository program would adversely affect nuclear development—as demonstrated by the embarrassment of the AEC in its efforts to proceed with waste disposal in salt in the early 1970s. Yet because there is an existing inventory of wastes to be disposed of, such a failure would not necessarily impede SSD.

**For example, currently it seems unlikely that TRU wastes, which are much bulkier than HLW, could be economically emplaced under the seabed, and this option has not been studied. This means that a major terrestrial disposal system would have to operate parallel with SSD. There would be little environmental rationale to pursue both methods of disposal, but there could be a political rationale if HLW disposal were much more controversial than TRU waste disposal.

31

Table 5

GEOLOGIC WASTE ISOLATION RESEARCH ACTIVITIES BY COUNTRY OR AGENCY

Country or Agency	Waste Type	Repository Formations under Consideration	Research Activity Current Studies	Milestones
Austria	Spent fuels Miscellaneous wastes	Hard rock	Site evaluation Granite properties Safety assessment	
Belgium	HLW Non-HLW	Clay beds	Site evaluation Clay bed properties In situ studies—tunnel	Pilot repository at Mol for alpha wastes and non-HLW—1981
Canada	Spent fuels HLW Non-HLW	Bedded salt Plutonic rocks	Site evaluation In situ studies—granite test site	Commercial repository: Site selection—1981 Demonstration—1985 Startup—2000
CEC	HLW		Catalog of potential sites in member countries Support work in Belgium (clay), FRG (salt), United Kingdom and France (crystalline rock)	
Denmark	Non-HLW		Geologic survey	

Table 5 (cont.)

Country	Waste	Rock	Activities	Status
FRG	Spent fuels Non-HLW HLW	Salt	Salt mine operation— Assa mine Safety assessment Engineering for spent fuel and HLW storage	Commercial repository at Gorleben—late 1980s
France	Alpha wastes	Rock salt Crystalline rock	Geologic survey Safety assessment—Oklo[a]	Pilot plant repository (alpha wastes)—1985
GDR	Non-HLW	Salt	Repository development at Bartensleben	
India	HLW Non-HLW	Igneous rock sediments	Site evaluation Properties of rocks	
Iran			U.K. Atomic Energy Agency evaluation of disposal in Iran	
Ireland	Non-HLW		Geologic survey	
Italy	HLW	Clay Salt	Geologic survey In situ studies in clay beds	Pilot plant repository—mid-1980s
Japan	HLW	Varied rock	Geologic survey Safety assessment	
Netherlands	Non-HLW	Salt	Geologic survey Safety assessment	
Spain	HLW Non-HLW	Shale Clay Salt	Non-HLW going into old uranium mines Site evaluation	Pilot plant repository—mid-1980s

Table 5 (cont.)

Country or Agency	Waste Type	Repository Formations under Consideration	Research Activity	
			Current Studies	Milestones
Sweden	Spent fuels HLW	Hard rock (granite)	Field tests—Stripa mine Safety assessment	Pilot plant repository—mid-1980s
Switzerland	HLW Non-HLW	Anhydrite Clay	Site evaluation Safety assessment Development of test site for HLW	
United Kingdom		Clay Crystalline rock	Geologic survey Safety assessment Site evaluation	Commercial repository—2000
USSR	Non-HLW	Varied	Direct injection underground	

Source: U.S. DOE, *Nuclear Waste Management.*

[a]Oklo, a site in West Africa, is the location of a naturally occurring body of uranium ore in which nuclear reactions occurred spontaneously in the distant past. Oklo is a natural analogue to a waste repository, and provides one of the only geologic formations that permits estimates of the speed with which radioactivity in a geologic repository would migrate into the biosphere.

The backup status of SSD also means that funding for research and engineering will be limited by expenditures for the terrestrial program. This has not been a serious problem thus far for two reasons. First, SSD studies have concentrated largely on basic science, which is substantially less expensive than engineering. Second, the NWPA has created a new means for financing the development of terrestrial repositories—a tax on nuclear-generated electric power; this tax has for now eased funding pressure. As the SSD program moves from scientific feasibility to technological proof of concept, however, funding may become a constraint on both the timing and scope of the research effort.

More serious in the long run is the likelihood of overall constraints on spending for radioactive waste disposal. Such constraints would force program managers to balance the need to complete a high-cost first-generation option against the desirability of proceeding to develop a backup system like SSD. The experience of the National Aeronautics and Space Administration with the space shuttle in the 1970s suggests that desirable research is no match for imperative need when the budget tightens. The SDP has experienced the threat of budgetary constraints since its inception.*

Yet another important aspect of the second-generation character of SSD is that the on-land components of the system are likely to be similar to those developed for the first-generation option. The similarity is necessary not only for economic reasons, but also to provide technological compatibility between the waste-handling systems. This means that the marine components will be *constrained by choices already made* in the terrestrial technology, such as the design of shipping casks and the setting of occupational exposure standards. In addition, an international disposal technology like SSD must contend with the likelihood of incompatibilities between national systems. Clearly technological considerations point to the need for the development of common standards and the desirability of discussing these standards while the terrestrial system is still flexible.

---

*When the SSD concept was originally formulated in 1973, it was clear to Hollister and Anderson that the idea would be greeted skeptically in the nuclear establishment (Hollister, Belmont talk). Quiet but effective liaison with legislators and Congressional staff interested in energy and ocean policy has helped the program to weather budgetary turbulence.

SUMMARY: A COMPARISON OF TERRESTRIAL DISPOSAL AND SSD

Like the terrestrial disposal program, SSD is affected by its history; moreover, more research and development must be done before geologic isolation of radioactive wastes either on land or under the seabed can be determined to be acceptable and feasible. SSD differs from the terrestrial program in the United States and elsewhere because of its distinctive intellectual and bureaucratic origins, and because its technical development follows in the wake of related work on land. Some major points of comparison can be made between the two systems:

1) Scientific base: Terrestrial disposal research has gradually, but often reluctantly, made use of the analytic capabilities of geology and mining technology. SSD began as a scientist's inspiration, and it has maintained its status as an intellectually exciting project pursued by elite scientists; this is one of the program's major strengths.

2) Technical acceptability: As with any undertaking, risks to human populations and the environment cannot be completely controlled. Since these risks are politically salient, however, any means of disposal will involve judgments of acceptable risk, made perhaps by default.

3) Reliability: The history of waste management in the United States has largely been one of institutional and political errors made in short-sighted attempts to avoid technical mistakes or to accelerate programs. The reliability of first-generation geologic isolation would be difficult to assure even without the controversy surrounding nuclear power. Although SSD can benefit substantially from its second-generation status, the reliability of the waste management system will be under close scrutiny because geologic isolation under the seabed will be a first-generation technology in the international arena. That the domestic elements of SSD are likely to have to meet tighter constraints than land-based programs may be helpful in meeting the test of reliability.

4) Legitimacy: A major issue in the development of terrestrial repositories has been jurisdictional conflict. In the Federal

Republic of Germany, a proposed salt repository near the town of Gorleben was scuttled by opposition from the state. In the United States the Waste Isolation Pilot Project has been plagued by troubled relations between the state of New Mexico and the federal government. The degree to which state governments can interdict national programs has been a principal issue in the passage of legislation for waste management. Similar issues may arise with respect to SSD, affected by the world ocean regime created under UNCLOS III.

5) Liability and compensation: U.S. policies for radiological injuries are largely framed by the Price-Anderson Act, a law which limits the compensation that can be claimed by those injured. One feature of SSD that is worth further investigation is the effect of shifting tort judgments from national to international arenas. In the case of both terrestrial disposal and SSD, however, the probability that no damages will appear for a long time after exposure complicates findings of fault.

6) Urgency: As argued above, a major institutional difference between terrestrial disposal and SSD lies in the controversy and urgency attached to the first repository. If SSD turns out to be the best option for a particular nation, that nation is likely to find itself paying most of the bills for research and engineering. It is not probable that the United States—which is now the major backer—will find itself in that position.

From a technological perspective, SSD clearly shares many of the characteristics of geologic disposal on land. However, SSD technology creates two important institutional conditions: (1) the need for international cooperation and supervision in research and implementation; and (2) the need for distinctive interactions with terrestrial research programs because of SSD's second-generation character.

Chapter 4

POTENTIAL ORGANIZATIONAL AND OPERATIONAL PROBLEMS OF THE SSD OPTION

The organizational and operational problems which may be posed by the creation of an SSD system for HLW have not yet been determined because all the technical/engineering details of such a system are not yet clear, nor has work on the policy issues of this dimension gone beyond the rudimentary posing of significant questions. In this chapter we shall focus on the land-based and sea-based components for system design and management and what they imply for effective and reliable operation of an SSD system. We shall also deal with the special problems of monitoring and managing the complexity which an SSD system is likely to generate.

SYSTEM COMPONENTS AND REQUIREMENTS

Land-based components of an SSD system will include the following: transformation of the waste form; handling and transportation to a port facility; interim storage, including shielding and cooling; and methods of transfer to disposal vessels.[1] Under current law these elements are regulated by the U.S. NRC.[2] The fact that a system for terrestrial disposal will have been designed—and probably implemented—before an SSD system means that the regulatory regime is likely to be patterned on the one developed already. In particular, the National Environment Policy Act (1969) process for developing an environmental impact statement would apply.

Waste form transformation and handling and transportation will probably be required to meet the same performance standards as in a terrestrial system. Transport casks for on-land handling of HLW will probably be the same as in the terrestrial system—an outcome which may constrain the design of SSD canisters. However, an additional requirement is likely to be placed on SSD canisters so that they are able to withstand the extreme pressures encountered during emplacement in the deep sea.

Port facilities are similar to Away-From-Reactor (AFR) storage facilities from a regulatory standpoint. However, within the existing framework of regulation the former have several possible points of vulnerability. For one thing, the high population density in port cities could raise difficulties of complying with established exposure standards. Moreover, port facilities raise questions of overlapping jurisdictions. The authority of the NRC to regulate installations of the DOE is not clear-cut. In addition, the loading of ships with radioactive waste—technologically similar to the handling of spent nuclear fuel in a reactor—may entail regulation by the U.S. Coast Guard.

The regulatory issues raised in the United States have counterparts in other nations participating in SSD research. Designing a multinational disposal program must take account of the variety in national regulatory regimes. The costs of separate and incompatible technologies need to be balanced against the difficulties of ensuring compatibility throughout the long period in which both first- and second-generation waste handling systems will be developed.

Several points need to be emphasized with regard to the land-based components noted above. First, only HLW in a liquid form seems to be appropriate for SSD because it can be solidified into a borosilicate glass waste form.* Spent fuel assemblies from light water reactors should not be disposed of in the sub-seabed because the first criterion of the multiple barrier concept cannot be met. If these assemblies are reprocessed on a large scale, significant liquid HLW will be produced which can be transformed into a waste form appropriate for SSD (presumably at the storage site). As pointed out in Chapter 3, however, the economics of commercial reprocessing of spent fuel are adverse. At present in the United States it is only the HLW from military uses and the commercial reprocessing of civilian spent fuel which could be disposed of in the sub-seabed. Thus we have characterized SSD as a second-generation (or later) technology.

---

*As noted in Chapter 2, the intensely radioactive liquid that is left after chemical separation or reprocessing of spent reactor fuel is called high-level waste. Only transformed solid wastes are suitable for SSD because they meet the criteria of the multiple barrier concept.

Second, the only differences between the land-based components of a terrestrial disposal system and an SSD system lie in the requirement of a port facility for the latter, the choice of transportation routes, and the methods of transfer to the transport ship. The port facility requirement will raise three sets of management issues: (1) Site location and cost ceilings which would secure local and state government approval. Many of the potential problems connected with this issue could be avoided if military facilities were used. (2) Selection of transportation routes. Like the first issue, this one will be politically sensitive as well. (3) Capacity of an SSD system to handle varying loads and the rates at which wastes could be disposed of. This issue raises a secondary one on the number of disposal ships that will be operating.

The sea-based components of an SSD system will include engineering design problems which remain to be solved given the assessments of risk, the requirements for salvage in the event of accident, and the emplacement technology.

The Sandia Laboratories SDP divides risks into two categories: predisposal and long-term or postdisposal. Predisposal risks include all components of the land-based system as well as risks attendant on sea transportation activities.[3] While preliminary work has been done in both categories, it has been only exploratory. For instance, James D. McClure has analyzed the probability of spent fuel transportation accidents in the United States and shown that it is quite low (5.0×10^{-7} accidents/mile).[4] The System Analysis Task Group of the NEA Seabed Working Group has assessed the risks of operational failures involved in transporting wastes on ship to a disposal site and their emplacement in the disposal medium.[5]

Likely accident scenarios will have to include ships being sunk in highly adverse weather conditions, accidents in congested areas, and accidents occurring during emplacement. Thus from a management point of view the design of the vessels, the choice of routes, and the emplacement technology will be important.

The alternatives currently being discussed for SSD emplacement are the "penetrator" concept and the "drilled-hole" concept. The former covers two technologies: "those characterized by a free-fall mode . . . and those requiring boosting . . . using either conventional propellants or the ambient hydrostatic pressure."[6] The latter would be a multiphase operation:

(1) establish the emplacement hole, (2) emplace a canister at the end of the drill pipe, (3) withdraw the drill pipe a specified distance, (4) backfill with local material and (5) repeat steps two, three, and four the required number of times for the specific hole location, (6) withdraw, backfill, and reconsolidate the sediments above the last canister. The emplacement site would then be left with a re-entry cone and any other required monuments [marks of location] and instrumentation.[7]

Different risks might apply to each alternative, raising two questions: (1) What is the optimum disposal fleet size given the amount of HLW to be disposed of, the location of disposal sites, and the available choice of routes? and (2) Is it necessary and/or desirable to standardize disposal technology internationally? We cannot now answer the first question because the necessary information is not yet available, but given the total amount of HLW existing in a form suitable for the SSD option, it is possible to estimate the optimum fleet size in the same way that it is possible to estimate the total number of repositories.

International standardization of disposal technology, defined primarily in terms of vessel design and emplacement technology, would substantially increase the complexity of the management problem because standardization across national boundaries and jurisdictions would have to be negotiated. There would have to be trade-offs between high levels of complexity and system reliability. A priori there does not appear to be any reason why this kind of standardization would be necessary since multinational use of a single site would be possible without standardization provided that all vessels could navigate and position themselves accurately at the site and provided that the nations involved could cooperate to allocate space within the site. (However, international reliability criteria may well be higher for the SSD option because accidents prior to emplacement could mean that unless canisters can be retrieved, they would be in the biosphere.)

DESIGN OF THE MANAGEMENT SYSTEM

On the basis of the technical characteristics described above, the management system for SSD must be designed in terms of a serially interdependent (or long-linked) set of techologies.[8] Such interdependencies operate in two ways. First, if one part fails, it jeopardizes the other parts; second, the outputs of preceding components act as inputs to those that follow.[9] Serially interdependent technologies tend to generate vertically integrated management systems in order to reduce uncertainty, but problems of coordination are magnified when they contain both types of interdependencies because each contains increasing degrees of contingency.[10]

Additional characteristics, unique to SSD, could make the system even more difficult to manage effectively. These have been illuminated by Todd LaPorte. The most important is that the management system must be super-reliable—i.e., it must assure "nearly error-free operational management, as well as nearly escape-proof burial, of radioactive wastes."[11] (LaPorte defines error as "any release (escape) of radioactive materials from the operational system such that recapture is either impossible or too costly to effect.")[12]

We assume that the primary goals of an SSD system will be twofold: the reliable disposal of wastes and the protection of the marine environment and man. However, if in response to political pressure exerted by publics fearful of any error, the official design criterion is to produce error-free management systems, the result will be a demand for "decision-making without feedback."[13] In itself, this is an extraordinary demand. Placed in a context of large, highly complex, and routinized systems, it is a challenge which we think is extremely difficult to meet because it excludes the possibility (and flexibility) of the system as a whole developing by trial and error or incrementally.[14]

When engineering design alternatives become clearer, research on alternative configurations of management system design at the national level will be necessary. The research agenda has already been specified by LaPorte:

1. Studies of the time periods involved, the geographical dispersion of activities, the types of skills required, the complexity of internal coordination, and the number of administrative functions in support of operational activities.

2. Estimates of resources (manpower, financial, and logistical) required to perform each function and the probable social impacts generated.

3. Studies of the "operational reliability of sociotechnical systems as their scale and internal complexity increases."[15]

The Battelle Human Affairs Research Center has made a start on the detailed conceptualization of some of the management issues.[16] With respect to the issue of time, it identifies three periods of institutional management based on the thermal and radioactive generating capacity of the isotopes involved: years 1-100, years 100-700, and years 700-negligible toxicity. Battelle suggests that three questions particularly germane to the issue of time need detailed consideration:

1. Under what conditions can we have any assurance that the societal institutions will last long enough to carry out the necessary management activities to insure the safety and integrity of waste repositories?
2. Under what conditions will the institutions set up to manage the waste repositories be durable enough to withstand changes in political regimes or other societal changes?
3. Under what conditions will the institutions set up for waste management operate as they are supposed to? Will they be competent? How will such institutions be regulated or controlled?[17]

There can be no final answers to these questions before the fact. While there are many examples of organizations operating successfully for 100 years and even some operating for 200-700 years, it would be folly to suggest that we know how to design organizations to last for those periods of time. Once organizations are created, they take on lives of their own and often develop in ways quite unintended by their creators. Organizational dynamics tend to revolve around a search for autonomy, security, and prestige; these in turn involve competition for resources, "turf," and political support. The organizations interact with their environments in ways that both can be foreseen and are unknowable. While it is clearly possible to maintain societal support for the performance of certain tasks for 100 years or more, time in itself can be an independent variable, and combinations that are possible at one point may not be possible at another.

In addition to the uncertainties relating to long-term institutional management, problems of risk and system reliability will arise from two sources: the technical operations themselves, and the performance of the management system and the meshing of the technical operations with the management system. The major problems in the latter category will stem from the very high levels of internal complexity that will be generated as a result of the large number of system components, their variety and technical sophistication, and the high level of interdependence among them.[18]

The system will be required to perform the following: operations, control and management, monitoring, and information transfer.[19] We have already sketched the operations and control and management functions. The monitoring function is itself complex because it will require the following activities:

1. Testing and pre-operational engineering development.
2. Pre-operational site characterization.
3. Monitoring of the politically sensitive transportation system.
4. Route position monitoring.
5. Surveillance during scheduled operations.
6. Routine post-operations surveillance.
7. Emergency monitoring and procedures in the event of accident or system failure.[20]

The greater the degree of internal complexity, the higher the likelihood of coordination failure; therefore the greater the need for built-in redundancy. Furthermore, as noted, organizational design will tend toward vertical integration. While vertical integration will help to mesh operations with control and management, it will not reduce the risk of component failure to zero. Moreover, problems will remain in meshing national systems with regional and global regimes. Each attempt at coordination will add another layer of complexity to the total system.[21]

On the basis of the work done thus far, it is clear that the organizational component of operating a HLW disposal facility— whether terrestrial or sub-seabed—could be the source of major problems. The organizational design problem is novel, and the demands placed on the system for super-reliability and longevity are enormous. Moreover, two additional potential problems are unique to the SSD option. First, for reasons discussed above, SSD is at least a second-generation technology for the United States but not for some of its allies who lack terrestrial alternatives. Second, the management requirement of longevity far exceeds the capabilities of human beings to maintain social institutions which perform specific tasks. This implies that irretrievable disposal is the only sensible alternative. However, irretrievable disposal is much less politically acceptable (as we shall see below).

In addition to the organizational and operational issues we have discussed, the SSD option will be affected by sociopolitical expectations (as noted in Chapter 3). Two predictions have been made regarding possible effects. According to one, if SSD is a second-generation technology and if the terrestrial option works, the link between HLW disposal and the growth of nuclear power may well be broken for the SSD option. In these circumstances the burden of the super-reliability criterion for SSD may also be reduced. According to the alternative prediction—as argued by LaPorte, for example—the link will not be broken because public perception of the SSD option will be shaped by the very different technology involved and by concern over how much HLW the sub-seabed can accommodate. The differences in terrestrial and SSD technologies are sufficiently great that the SSD option could not escape the super-reliability criterion. The question of sub-seabed accommodation capacity is not just technical but includes a substantial amount of political perception and judgment; moreover (as noted), large-scale application of SSD implies a prior commitment to commercial reprocessing of spent fuel.[22] If this prediction is correct, the future for the SSD option will be more uncertain.

Chapter 5

POTENTIAL DOMESTIC POLITICAL PROBLEMS
CONNECTED WITH THE SSD OPTION

The complexity of the U.S. political system makes analysis of present political problem development difficult. To predict potential problems at least two decades in advance is even more difficult, and analysis becomes more perplexing when one considers the scientific and theoretical complexity of radioactive waste management and SSD as a political issue. Nonetheless, a number of observations and tentative conclusions can be made about the domestic political problems which are likely to be encountered in the development of an SSD program.

Our analysis in this chapter will focus on three major considerations:

1. The political issues concerning SSD and a comparison with similar issues in other problem areas;
2. The most probable political actors and their interests and strategies;
3. External factors likely to affect both the SSD option and the actors.[1]

After presenting the issues, actors, and external factors in detail, we shall analyze them together in an effort to assess their impact on SSD.

POLITICAL ISSUES OF SSD: SUBSTANTIVE AND TACTICAL

If the United States were to proceed, or assist its allies to proceed, with the development of an SSD program, it would arouse two broad types of concerns: substantive and tactical. (The effects of the two will be considered in the last section of this chapter.)

46

Substantive concerns would focus on two main areas: nuclear waste management and the use and environmental protection of the oceans. These concerns may include ethical, legal, social, political, or technical dimensions.

Predominant concerns in nuclear waste management have been identified by limited research on nontechnical issues.[2] More specific to the SSD issue are concerns over HLW repository siting. Considering dominant American values, Ted Peters has examined four major issues related to repository siting: uncertainty and risk, geographic equity, intergenerational responsibility, and implementation ethics.[3] Under uncertainty Peters includes the effects on health of low levels of radiation over long time periods, whether technology exists to prevent toxic wastes from escaping to the biosphere, and human fallibility and malevolence.[4] The issue of risk involves the quantification, perception, and comparison of risks, the distribution of such risks between individuals and societal groups, and risk mitigation. Geographic equity involves the perception that people living near a repository will be involuntarily subjected to un-deserved risk, whether or not they benefit directly from the waste-producing technology. Intergenerational responsibility concerns the impact of nuclear waste disposal on future generations and the question of whether present generations have the right to bequeath risks and responsibilities. Implementation ethics deals with whether the implementation of a disposal program could be efficiently and safely managed and whether public acceptance would be required.

The second major substantive concern, the use and environmental protection of the oceans, stems from a widespread belief in the United States that the oceans are important—if not essential. Ocean usage issues include the need for wise management of ocean resources, equitable access to resource exploitation, and legal and policy aspects of ocean use (including law of the sea questions and concern over the fragmentation of ocean policy and regulation). The protection issue involves preservation of the ocean environment, conservation of living marine resources, and protection of marine mammals and shoreline habitat. In addition to present interests, many of the ocean issues concern the ability of future generations to use, benefit from, and enjoy the marine environment.

While the issues discussed above arise in connection with nuclear waste management and the use and environmental protection of the oceans, the following concerns appear to be unique to the SSD concept: short- and long-term effects of temperature and radiation on the deep-sea environment, the adequacy of the sediments (as part of the multi-barrier approach) to prevent escape of radionuclides, potential pathways of released radionuclides back to humans, risks posed by accidents during ocean transport or emplacement of wastes, the organizational and operational requirements of an SSD program, the distribution of risks and benefits in the development of port facilities, and the question of whether use of the seabed for nuclear waste disposal will sometime in the future preclude other ocean uses deemed necessary or desirable.

In addition to substantive concerns, tactical concerns may affect the visibility and salience of the SSD issue. Factors motivating the key actors could be political, economic, or legal. At the political decision-makers' bargaining table, the SSD issue could be traded according to its perceived importance in relation to other current "hot" political issues. Researchers on the SSD concept have an interest in the issue's political salience because controversy can enhance or harm its funding (i.e., economic) potential. (Clearly those whose income is diminished because of allocations made for SSD—such as researchers on mined repositories—have an interest in the concept losing importance.) Finally, the implications of the SSD concept may provoke legal concerns over issues connected to SSD. For example, a political controversy over the legality of the SSD concept may affect perceptions of the legality of LLW dumping at sea, or attempts to amend domestic statutes to allow for SSD may upset an entire legal regime for protection of the marine environment.

While the SSD issue may share characteristics with many environmental and technology-related issues, it may be useful here to juxtapose it to an issue that seems exceptionally similar—that of ocean dumping of LLWs. Because of its similarity with the SSD issue and its apparent growing political importance (especially at the regional level), this issue may have important predictive value in our analysis below. In both issues concerns focus on the disposal of nuclear wastes in the deep ocean environment. Among the common concerns are short- and long-term environmental effects, potential pathways for released radionuclides to reach humans, institutional/

managerial problems (including the need for long-term monitoring), and equitable distribution of risks and benefits. At the same time, we must emphasize two important differences. The first is between high- and low-level wastes, and in particular their potential threats to the environment and to public health and safety. The second is between "dumping" (of LLW) and "emplacement" (of HLW).

PROBABLE POLITICAL ACTORS: IDENTIFICATION, INTERESTS, AND STRATEGIES

For the purposes of our analysis we shall both discuss five major categories of actors—the environmental and public-interest community; the nuclear industry and commercial interest groups; government; the scientific community; and the public—and subdivide them according to their interests and strategies.

ENVIRONMENTAL AND PUBLIC-INTEREST COMMUNITY

There is a wide variety of interests within the environmental and public-interest community, ranging from encouragement of SSD research to objection to the concept to linking SSD to what are considered more important goals, such as the cessation of nuclear energy development. In no case has this community promoted the SSD concept over alternative disposal options. It views SSD as futuristic and gives it a relatively low priority at this time, but it is well aware of both its technical and political aspects. In general, the community's concerns are procedural and focus on participation in decision-making.

One sector in this category comprises organizations that favor continued research on SSD but are concerned that the present research efforts of the U.S. SDP are less than objective.* Among the reasons for concern they include the following: (1) They allege a blurring of procedural steps that may hinder accountability and review. (2) Legal and institutional studies are being conducted or supervised by the same research organization (Sandia) responsible for sub-seabed technical research; as a result, Sandia is seen as a

---

*The SDP is described in the footnote on p. 21 above.

strong proponent of the SSD concept. (3) Funding cutbacks may negatively affect the SDP. (4) Criteria are needed, such as for costs and environmental safety, with which to compare land versus ocean disposal options. (5) Because pressure is building to solve the waste problem, bad decisions may result from moving too fast and not waiting for additional tests and evaluation. In summary, this sector is willing to wait and see what scientific studies determine and clearly opposes proceeding with a sub-seabed repository until knowledge is obtained and can be used in a comprehensive comparison of land and sea options.

A second sector in this category opposes the SSD concept altogether. Its primary interests are the protection and preservation of the ocean environment and its living resources, and it considers the SSD option less controllable than land repositories would be. Monitoring and constant scrutiny would be more difficult for logistical reasons and because of our lack of scientific knowledge and understanding of the deep-sea environment. In addition, it perceives the area at risk as much greater than that posed by land sites. Such scientific and technical concerns are mixed with deeply held beliefs that the marine environment has a special sanctity about it. (These beliefs are also held by a large segment of the public, and they will be discussed more fully below.)

A third sector consists of the anti-nuclear organizations. They believe that nuclear energy is both economically and environmentally unsound—chiefly because the permanent waste disposal problem remains unsolved. Not only are they interested in delaying further development of commercial nuclear energy, but also they hope to shut down existing nuclear plants, prevent new plant construction or operation, and prevent AFR waste storage.

It is difficult to predict which of the environmental and public-interest organizations will acquire the most political persuasion, or what the total effect of all the sectors will be. The extent to which these sectors coordinate their efforts in regard to the SSD issue may be an important criterion in political problem development. (It will be considered below.) Another criterion may be the strategies they choose to affect decision-making—e.g., coalition-building (domestic and international), linkage of the SSD issue with other politically sensitive issues, intervention into the adjudicatory process of agencies, legal suits, education of the public, participation in advisory or informational meetings and hearings, and direct lobbying efforts.

NUCLEAR INDUSTRY AND COMMERCIAL INTEREST GROUPS

The concerns of the nuclear industry and commercial interest groups center around an urgent need for a waste disposal solution. While the groups are aware that a technically and politically credible solution is necessary, they believe that a solution should be quick and obvious and not necessarily technically optimal. The nuclear industry is faced with a growing political and public relations problem as it struggles to convince a variety of formidable actors (including the courts, the states, and the public) that a solution can be found. Although the industry desires a workable statutory solution, its legal and technical dependence on a federal infrastructure has caused it much frustration. Its aversion to delay may cause it to oppose SSD if opposition would lead more quickly to a mined repository. Thus it may consider that the monies delegated to the SDP detract from revenue for mined repository research and development. The nuclear industry has a powerful lobby (its chief political strategy)—including coalitions of mining and petroleum companies, utility companies, and labor organizations—which it may use advantageously in future political arenas.

GOVERNMENT

Potential actors in this category include federal, state, and local governments. Within the federal government, prominent actors will be the administration, the Congress, and a number of federal agencies.

The Reagan administration plans to move ahead with permanent HLW repository development. It deemphasizes technical problems, considers political opposition manageable, and is allegedly exerting pressure on certain agencies—in particular the DOE—to quicken their procedural pace. The administration's interest in solving the waste disposal problem appears to be linked to its interests in promoting nuclear energy (both domestically and abroad) and developing the breeder reactor and reprocessing capability, as well as to its concern about the dependence of U.S. allies on non-Western energy supplies.

While the current administration's interests are important in determining the direction of repository research and development, the interests of future administrations will be equally important, are likely to be very different, and are essentially impossible to predict.

Congressional interest in the SSD concept is documented in two sets of Congressional hearings.[5] In general, the average Congressman is focusing very little (if any) attention on the concept at present—in part because of a lack of information, including data on the international aspects of SSD. Members who have expressed interest fall into several sectors. One sector is fundamentally concerned with why SSD research is being funded by the government when its political feasibility is questionable and when domestic and international law make it illegal. The prevailing belief of this sector is that DOE should not be spending money on a concept that is presently illegal. A second sector is concerned about past radioactive waste dumping at sea and has pushed for research on dump sites off New Jersey and in Massachusetts Bay. A third sector promoted passage of the MPRSA and ratification of the London Convention—statutes which allegedly make the SSD concept illegal. MPRSA prohibits the issuance of permits for ocean disposal of HLW.[6] The London Convention prohibits dumping—defined as "any deliberate disposal at sea"—of HLW.[7] This sector may become politically involved in the SSD issue if attempts are made to amend or by-pass these statutes. A fourth important sector is the strong anti-nuclear contingent within Congress.

A number of federal agencies are currently involved, or would be required to become involved, in development of a sub-seabed repository. Prominent among these are the DOE, the Environmental Protection Agency (EPA), the NRC, the Department of State, and the National Oceanic and Atmospheric Administration (NOAA). At least three of these have questioned the appropriateness of federal funding for sub-seabed research because of SSD's alleged illegality.

The DOE is an especially important actor; for the purposes of this analysis, it can be divided into two factions. The larger faction is involved with mined repository research and development and is expected to make a commitment to this disposal option. Any consideration of alternatives to the mined option would raise doubts about the adequacy of the DOE's present approach, so this faction considers that there is no need for an SSD option. Furthermore, this faction is interested in using DOE reservations for repository sites. The second faction is interested in continuing SSD research and development. By pushing SSD as a supplement rather than an alternative, it hopes to keep this option alive.

The EPA's primary official interest is in the potential for and effects of radionuclide release into the ocean environment. The agency is a potentially important actor in at least two respects. First, it is currently preparing criteria for deep geological disposal of HLW which could influence criteria for SSD. (It has not developed criteria for SSD alone because the agency views the option as illegal under both MPRSA and the London Convention.) Second, if the United States developed an SSD program, the EPA would likely be assigned the important task of preparing standards and permit procedures.

The NRC's role would be to implement regulations using EPA standards. Interested in the assurance of public safety and recognizing the limitations of models, the NRC is addressing the need for back-up systems in radioactive waste disposal. It considers that each separate component of a disposal system should be assigned a separate standard. This viewpoint is in contrast to that of the EPA, whose standards for deep geological disposal are being developed to apply to the entire repository system—an approach preferred by DOE.

The Department of State has an official interest in fulfilling U.S. obligations under the London Convention and maintaining the credibility of the international regime it establishes. It may therefore be required to assess the legality of SSD under the terms of the convention and the international implications of decisions regarding the SSD concept. In addition, the agency has responsibilities in several potentially related areas, including ocean disposal of LLW, proposed island storage of HLW, the radioactive waste disposal needs of U.S. allies, and nuclear energy promotion.

If the SSD option were developed, the NOAA would be required to collect data and assess the potential environmental impact of the program—including the impact on nonutilized or underutilized fish species, highly migratory fish species, and marine mammals. In addition, it would consider how estuarine or other coastal habitats would be affected by port development.

State governments are now and will continue to be important actors. Interests will vary between and even within states on the desirability of hosting a repository, establishing waste transport routes, or developing port facilities for handling wastes. State actors will affect the feasibility of mined repositories and thus indirectly affect other disposal options. In addition, state government reactions to ocean dumping of LLW and fuserap (contaminated soil) or scuttling of submarines may affect the SSD concept. California, South Carolina, and Hawaii are currently vocal actors in these issues.

Local governments may affect SSD in at least two ways. First will be their response to siting and development of port facilities for an SSD program. Second, they may indirectly affect decisions on SSD through their support or opposition to mined repository development. Let us emphasize, however, that in the case of both state and local government actors, the possibility of federal preemption should not be overlooked.

SCIENTIFIC COMMUNITY

Various sectors of the scientific community may become political actors in an SSD issue. In an effort to obtain funding, researchers may develop vested interests in particular disposal technologies. In this way the scientific community could affect both the level of technical research and the political visibility of the SSD concept. In addition, this group of actors could strongly affect public perception of the reliability of SSD and thus its feasibility. The public would rely most heavily on the opinion of the NAS and prestigious university scientists and to a lesser extent on government research and contract research organizations.

THE PUBLIC

The category of public actors refers primarily to the "attentive" public—i.e., the sector of the U.S. population likely to become politically involved in an SSD decision. A wide diversity of concerns will influence the level of public awareness of the issue. Currently the public is aware (at least regionally) of the issue of LLW dumping in the oceans. It is not clear that it will distinguish between HLW and LLW given the limited understanding of technical issues. However, it is clear that the public is concerned about the oceans and that there is a widespread belief that we cannot afford to poison them. It is also clear that public opposition is growing toward nuclear energy, partially based on an increasing concern that there will be a loss of human control over nuclear materials. This concern is heightened by the public's fear of radiation.

Concern for the ocean environment, although far from unanimous, not only appears quite pervasive, but also elicits strong emotional responses. Even many technically trained people – or those who have never lived near or seen an ocean – appear to react emotionally when ocean disposal of radioactive wastes is suggested. While the oceans appear to have a much larger public constituency than do land disposal sites, this constituency appears more diffuse and less vehement than land-site constituencies, with the exception of the oceanic environmental organizations.

FACTORS LIKELY TO AFFECT OPTIONS AND ACTORS

Factors likely to affect waste disposal options and probable political actors could be scientific, technical, political, or economic. With scientific and technical factors both the facts themselves and people's *perceptions* of the facts become important. For purposes of this study, we shall categorize factors into two major groups: (1) those affecting the success or failure of the mined repository disposal option and thus indirectly affecting SSD, and (2) those directly affecting people's perceptions of the feasibility of SSD.

Despite the fact that proponents of the SSD concept are presenting it as a supplement rather than an alternative to mined repositories, many potential actors view it as an alternative. Thus factors specific to mined repositories include technical adequacy of mined repositories; their general political acceptability; federal-state relations in siting of repositories; and local opposition to siting decisions.

Factors affecting people's perceptions of the feasibility of SSD will include the following: the scientific and technical adequacy of an SSD repository; monitoring and retrievability capabilities of SSD; geographic siting of repositories (including whether sites would be in the Pacific, Atlantic, or both; location of storage and port facilities within the United States; and whether sites would be located within or outside of the U.S. 200-mile Fisheries Conservation Zone—FCZ); cost-effectiveness of SSD (especially in comparison with other options); types of wastes to be deposited in the sub-seabed (commercial or defense-related; U.S. or imported); the degree of federal control over the disposal program; the extent and method of public involvement; conflicts over the use of the deep seabed; the extent to which the U.S. administration promotes nuclear energy; the urgency of permanent waste disposal needs (both those of the United States and its allies); whether the United States or some other country will lead SSD development; the current scientific assessment of LLW dumping in the oceans; the degree of coalition building; and the degree to which the SSD issue is linked with other politically sensitive issues. The linkage of SSD to other issues is complex. The most important linkage issues include nuclear energy development, LLW dumping in the oceans (especially in the Pacific), island storage of HLW, other ocean dumping issues, nuclear proliferation, nuclear disarmament, nuclear materials reprocessing, land disposal of HLW, scuttling of nuclear submarines in the oceans, preservation of marine mammals, and the common heritage principle for use of the seabed.

ISSUES, ACTORS, AND FACTORS: AN ANALYSIS

As should be clear from our discussion above, the complexity of the SSD issue, the large number of potential actors, and the variety of external factors could each create domestic political problems. In this section we shall attempt to analyze their characteristics and possible extent.

Several characteristics of the SSD issue appear to have political implications. First, the issue will transcend state and national boundaries. International concerns about the ocean environment and solution of the radioactive waste problem will result in a large number and variety of actors, increase the potential for coalition-building, and make political participation more effective. Strong and effective coalitions have been built in response to similar issues—for example, against nuclear energy, for environmental protection, and for the preservation of marine mammals. In addition, like these issues, the SSD issue may attain a high degree of national visibility and endurance. (One of the most predominant trends ever documented by the Harris public opinion polls involves the support for air quality and the Clean Air Act.)[8]

Second, SSD will be affected by the ocean dumping of LLW, an issue whose political importance is clearly growing. This issue encompasses U.S. plans to resume dumping radioactive materials (packaged LLW, fuserap, and decommissioned nuclear submarines) into the oceans, other nations' dumping practices in the Atlantic, and Japan's plans to dump LLW in the Pacific. In California, in reaction to U.S. dumping plans, the Senate Select Committee on Coastal Fisheries and Aquaculture has introduced a Senate Joint Resolution requesting Congress and the President to ban all radioactive waste disposal in California waters.[9] Moreover, the committee has proposed an international treaty to ban disposal of radioactive wastes in the Pacific. The committee has called for these measures until and unless future valid and reliable scientific studies prove LLW dumping to be safe. In Hawaii a U.S. senator has called for a U.S. policy against nuclear dumping in the Pacific.[10] Proponents of ocean dumping argue that no environmental or public health hazards have been documented by scientific research to date.[11] However, environmental public-interest groups have become actively involved, and an intense debate over the potential hazards of LLW ocean dumping is predicted.[12] If the United States resumes such dumping, a major political battle is expected.[13]

In order for SSD to attain national political significance, it will have to meet a number of criteria via the actors. We consider the first criterion to be continued government funding. Here the DOE is especially important. If the smaller faction within DOE that favors continued SSD research is unable to maintain funding for the SDP, the concept may lose political visibility. This would affect not only the U.S. program, but also an international program because the United States contributes approximately one quarter to one half of the total SWG research funds. With government funding, even if the functions of DOE are transferred to another agency (such as the Department of Commerce), it is likely that the HLW program (of which the SDP is a part) will remain intact.

The second criterion is a favorable Department of State assessment of SSD because the department brings international concerns into the national political arena. If in assessing the international

national ramifications of an SSD program, it decides that the United States should not lead and/or participate in the development of a sub-seabed repository (and so advises the President), it may prevent the issue from attaining national visibility.

The third criterion will be Congressional interest in the SSD concept. It will be the role of Congress to either promote or oppose SSD, depending in part on whether it considers SSD a serious or necessary disposal option and in part on how the issue will affect individual Congressmen politically. Congressmen may decide to promote SSD in an effort to dispel state and/or local opposition to mined repositories (and corresponding federal/state relations problems)—although it is not obvious that state and local governments will object to repository development. New jobs and potential increases in state and local revenues may overshadow perceived costs. Moreover, indications are that only three states will be chosen to host HLW disposal facilities. Non-host states with accumulating wastes may favor regional mined repositories, and their Congressmen may therefore oppose SSD. Finally, Congress may see the SSD concept as only a supplement to mined repositories, in which case it may consider it unimportant or unnecessary and ignore it. If the concept fails to attract enough Congressional interest to spur at least some members to promote or oppose it, its political importance will be limited.

The fourth criterion will be the involvement of the environmental and public interest community. If the various sectors of this community work together to promote or oppose the SSD concept, they could prove to be powerful actors. However, if a split arises between different sectors (for example, between oceanic organizations who oppose SSD and land-based organizations who promote it in lieu of land disposal), their political power would diminish. This community's concern about the SSD issue will increase if attempts are made to weaken MPRSA in order to legitimize SSD. Also important to these actors would be U.S.-approved changes in the London Convention and the U.S. role as protector of the marine environment. If the environmental and public-interest community did not become involved in the SSD issue, public and governmental decision-makers would assume its potential impacts were unimportant and significant political problems could not be expected.

The fifth criterion will be a lack of opposition from the nuclear industry. If the industry comes to view the sub-seabed concept as a major threat to its interests, it may oppose it. Such opposition could prove devastating to the political acceptance of SSD as an option deserving serious consideration.

The sixth criterion will be a lack of opposition from the scientific community. Like the nuclear industry, this community could greatly harm—if not destroy—the SSD option by publicly opposing it. For example, if the NAS or prominent scientists involved in SSD research determined that the concept was scientifically unreliable, it would no longer be seriously considered at the national level. While disagreement within the scientific community may enhance SSD's political salience, objection by a consensus or by certain sector(s) of the scientific community could prevent its politicization.

The roles of the EPA, NRC, and NOAA may be less critical but will add to the cumulative political problem development. The EPA's forthcoming criteria for HLW repositories may be unofficially used to judge the adequacy of the SSD option and may include such requirements as retrievability, which the SSD option may not be able to meet. Moreover, by maintaining its stance that SSD is illegal under both MPRSA and the London Convention and refusing to prepare standards for an SSD option, the EPA may hold up its development. The NRC, by emphasizing the need for a back-up system and pointing to the limitations of modeling (upon which the SSD concept is currently based) may affect perceptions of reliability. In addition, if the NRC is successful in its argument that each separate component of the disposal system must meet individual standards, it could greatly increase costs and make a supplemental sub-seabed repository far too expensive. Furthermore, the NOAA requires an environmental impact analysis for SSD. Such an analysis could arouse public controversy since the requirement implies the possibility of environmental hazards. If the NOAA analysis discovers possible hazards, political controversy will heighten.

Future administrations may affect national involvement in the SSD issue by their promotion of or opposition to increased nuclear energy development, reprocessing, and U.S. assistance to allies with limited land disposal options. If future administrations continue to promote the use of nuclear materials, the United States may decide to proceed with the SSD option. If they oppose further development of nuclear energy and reprocessing, the United States will probably not be interested in the SSD option unless mined repositories fail. However, future administrations may decide that the United States should perform an oversight or regulatory function if other countries proceed with sub-seabed repository development.

All of the above actors will contribute to the public's perception of the SSD concept. It is difficult to predict how important public opinion will be because it hinges on two generic policy questions: (1) To what extent is the public able to affect the political process? (2) How will more information change the public's views? Nonetheless, the extent and method of public involvement may affect the political salience of the SSD issue.

Several factors are likely to affect political problem development. First is the matter of whether SSD is viewed as an alternative or a supplement to mined repositories. If seen as an alternative, the concept may receive greater Congressional, state, and/or public support. However, it may have to be viewed as a supplement to continue to receive DOE funding and escape nuclear industry opposition.

Second, because SSD will be viewed by many as an alternative, the status of mined repository development will be an important factor—i.e., its technical adequacy, political acceptability, and whether it becomes necessary to take pressure off the states who would be potential hosts for land disposal sites. If mined repositories are deemed unsuccessful, the SSD option may appear more acceptable, and less political problem development could be expected. If, on the contrary, mined disposal proceeds as planned with reasonable success, SSD may be viewed as an unnecessary risk.

Third, the scientific reliability of SSD remains to be determined. For one thing, the SDP is still in the early stages of analyzing reliability. For another, there is no reason to believe that there will be a consensus within the scientific community. Some questions related to SSD are highly controversial—for example, the health effects of long-term exposure to low levels of radiation. Disagreement over the scientific aspects of SSD within the scientific community may add to the public's mistrust and apprehension about the entire radioactive waste problem. Moreover, a calculation of risks to the environment and to public health and safety involves the generic policy problems of how much information is enough and with whom the burden of proof or persuasion should lie. If it is up to the proponents of SSD to prove its safety, the option may never prove politically acceptable. On the other hand, if its opponents must prove it unsafe, its chances for serious consideration will be enhanced. Judicial decisions in assigning the burden of proof may prove pivotal if scientific debate over SSD reliability is aired in the courts.[14]

Opinions within the scientific community will be influenced by another factor—i.e., whether monitoring and retrievability will be possible with SSD. Politically, assurance of monitoring capability is important because monitoring in the deep ocean is generally considered more difficult than mined repository monitoring. Retrievability adds a safety back-up if problems occur or if a better disposal technology is discovered through future research. Providing for retrievability could also allow SSD to be defined as "storage" rather than "disposal," a distinction that has many legal implications. However, retrievability could be considered a disadvantage if it were perceived to increase proliferation risks.

Several geographic factors are likely to affect political problem development: (1) Will disposal site(s) be in the Atlantic or Pacific? For several reasons fewer political problems are likely if an Atlantic site is chosen: there is no history of nuclear testing (and resultant human tragedy) in the Atlantic; there is no political movement in the Atlantic akin to the "Nuclear Free Pacific" movement; and current LLW dumping is being carried out in the Atlantic by many of the potential users of SSD. (2) Will storage and port facilities for an SSD program be built on the east or west coast? Because more wastes are generated in the eastern United States, more political problems may be encountered if these wastes are transcontinentally shipped to west coast facilities. (3) Will site(s) be located within or outside the U.S. 200-mile FCZ? More domestic political problems can be expected as sites are moved closer to shore. However, if international waters are used, the common heritage issue may add to political problem development.

The factor of cost-effectiveness will prove important if SSD is considered as an alternative to mined repositories and cost comparisons can be made. At present, cost estimates for SSD have not been developed. If they prove to be very high, SSD as a supplement may not be justifiable. On the other hand, safety implications may outweigh cost considerations.

Whether wastes are commercial or defense-related may prove

politically important. If a repository is designated for commercial wastes, it will more likely be linked with the nuclear energy debate. If it is designed for defense-related wastes, it may be linked with the growing nuclear disarmament movement. Yet many people will view a defense facility as a necessity, and less political controversy may ensue.

The extent and method of public involvement is a complex factor, potentially affecting public and governmental decision-making. The methods range from information dissemination to participation in governmental decision-making. The more information provided on the SSD concept, the more questions will be raised. As noted, it is not clear to what extent or in what way information will affect the public's decision. For example, when the public is provided risk calculations, it tends to focus on the magnitude of the risks and discount their probability.[15] Research on public opinions about radiation indicates that the basis for most decisions on the subject is the way people feel about it.[16] This raises the question of whether information dissemination efforts are cost-effective. To the public, the amount of information may be less important than the truth of the information.[17] Only candid communication of information relevant to human safety and well-being will enhance public trust.[18] Perhaps the most effective method of involvement would be public participation in the decision-making process, resulting in more openness and better decisions. Information dissemination alone may increase political problem development.

One of the most important factors surrounding the SSD issue is the degree to which it is linked with other politically sensitive issues. Greater political problem development can be expected if it is linked with any of the following issues: nuclear energy development, nuclear disarmament, nuclear proliferation, nuclear materials reprocessing, land disposal of HLW, the common heritage principle for use of the seabed, ocean dumping of toxic wastes, ocean dumping of LLW, preservation of marine mammals, scuttling of nuclear submarines in oceans, and island storage of HLWs. The SSD concept has already been linked to LLW dumping, scuttling of nuclear submarines, and HLW storage on Pacific islands by the media,[19] public interest organizations,[20] and California Senate Joint Resolution No. 27.[21]

A final very important factor will be the degree of coalition-building among actors. The more there is, the greater will be the political leverage of the coalition groups, leading to an increase in the political problems. For example, if SSD is linked to the disarmament issue, environmental organizations may join forces with church and other disarmament groups. Other potential coalitions include the environmental community and Third World organizations (over appropriate use of the seabed); health physicists/physician groups with the environmental community (over potential public health hazards); industry and environmental groups (opposing the DOE); domestic and international environmental organizations; anti-nuclear groups with the environmental and public-interest community (in efforts to halt nuclear energy); and state and local organizations with national public-interest organizations (opposing mined repositories).

There appears to be a great likelihood for political problem development if the United States decides to proceed with SSD. Even if other countries proceed first, some domestic political reaction should be expected because the U.S. role as protector of the marine environment and of public health and safety would be at issue. Therefore, for domestic political reasons the United States should expect to be required to fulfill at least an oversight role if future SSD development occurs.

The United States perhaps faces greater domestic political difficulties than other interested countries because of legal impediments to SSD such as MPRSA. Moreover, the SSD issue will likely be most salient within the United States because this country has led in technical and nontechnical research on the SSD concept, because environmental and public-interest group opposition to nuclear development has been greatest here, and because many nations appear to be waiting for the United States, as the developer and promoter of nuclear power stations, to decide on a solution to the HLW disposal problem. The importance of political reactions within the United States to the overall outcome of the SSD option is difficult to measure. David Deese suggests the following:

> It is entirely possible that the United States may be forced to abandon part of the [SSD] program on international political or legal grounds. Even more likely is the probability of abandonment as a result of bureaucratic or environmental opposition in the United States.[22]

For purposes of this analysis, a useful way to measure the importance of domestic political problems to the eventual outcome of the SSD issue is in terms of their effects on international SSD problems. Many of the problems addressed in this chapter promise to add to international SSD problems. (We discuss these in Chapter 6 below.) For example, the way in which the United States assesses and deals with the various risk factors discussed above will affect the feasibility and reliability of the international SSD issue. Feasibility may also be affected by U.S. public opinion. We have noted that the U.S. public exhibits both concern about the ocean environment and mistrust and apprehension about the radioactive waste problem. These factors are likely to contribute to a global preference for caution. U.S. legal and political determinations on the applicability of the London Convention, common heritage principles, and UNCLOS III will influence the legitimacy component of the international issue. Whether U.S. political interests extend to assisting allies with their HLW disposal problems will affect the urgency factor.

Specific aspects of potential domestic political problems seem especially likely to shape the SSD issue in the international political arena. For example, a linkage of the ocean dumping of LLW and/or the anti-nuclear issue with the SSD issue in the United States may lead to such a linkage at the international level as well. In addition, the international character and membership of environmental organizations (including oceanic ones) and the transnational anti-nuclear movement may well be affected by their U.S. counterparts. Finally, in the design of an international sub-seabed regime, two important criteria, feasibility/acceptability and urgency (discussed in Chapter 6), will be decided primarily at the national level for countries possessing SSD capabilities.

Chapter 6

POTENTIAL INTERNATIONAL PROBLEMS
CONNECTED WITH THE SSD OPTION

The potential major international problems of an SSD option can be analyzed by considering the following: (1) The SSD issue from an international perspective; (2) International analogs to the SSD issue and links between SSD and related issues; and (3) Likely major problems of negotiating an SSD regime and how to approach them.

THE SSD ISSUE FROM AN INTERNATIONAL PERSPECTIVE

When we disaggregate the mix of potential problems connected with SSD at the international level, they appear to involve the following: feasibility, reliability, legitimacy, liability, and urgency.

FEASIBILITY

The question of whether particular technological systems are feasible may seem to imply that simple yes/no answers are possible, but in the case of SSD the answers are considerably complicated. For instance, feasibility will include judgments that go beyond whether it is possible to emplace canisters of vitrified HLW into the sediments of the sub-seabed and whether we can expect the multiple barrier system to function as intended. The essence of the feasibility judgment will be an assessment of levels of risk with respect to accidents, release rates, potential pathways of released radionuclides to man, possible harm to the marine environment, and possible harm to man. But estimates of risk pose issues of political acceptability, and there are different ways for the risk issue to be salient. Moreover, not only must political authorities consider risks acceptable, but also relevant publics must agree. In the international system relevant publics will include governments who cannot make independent judgments on the SSD option, as well as national nongovernmental and transnational groups (see below). Securing official and public agreement will be complicated by the need to make comparisons between terrestrial and sub-seabed options.

While the issues of terrestrial disposal are not different in kind from SSD, the levels of uncertainty and operational difficulty in the ocean are a great deal higher. At the same time, it must be recognized that for a few countries with large and growing stockpiles of HLW and little possibility of terrestrial disposal, the urgency will be greater, and these countries may consider SSD a first-generation option. In consequence, the international community will have to make judgments about the feasibility of the SSD option before most participants are ready to do so. We refer to the international community as a whole because we consider that substantial joint action will be necessary in the determination of feasibility. The determinations of the SWG along with independent assessments by countries possessing a nuclear disposal capability will not be sufficient.

RELIABILITY

Presumably the ultimate objectives of an SSD system are the safe, long-term disposal of HLW and protection of the marine environment. These objectives imply at least five sub-issues of considerable importance. First, what risks are associated with possible accidents in the different phases of an SSD system? There are five operational phases to be concerned about: packaging of the waste, transportation to a port facility, handling and loading on ship at the port, transportation to a disposal site, and emplacement of canisters. The first three phases will be of primarily national concern; the last two will be of primarily international concern. In addition, there will be significant national and international concern for definition of standards relative to ship design, construction, equipment, and manning. Second, what safeguards are available to reduce the risks or at least prevent them from increasing? Third, as noted, a significant monitoring capability will be required for an SSD system to be judged reliable. At the international level this poses two questions: Can canisters be monitored effectively once they have been emplaced? If so, can they be monitored for the time necessary given the half-lives of the isotopes involved? Fourth, can canisters be retrieved once they have been emplaced? Fifth, what are the likely effects of accidents? Even if the risks are judged to be low, the fifth sub-issue will probably be the stone on which the whole SSD concept will turn. Thus the fourth sub-issue is likely to be the critical variable in international judgments of system reliability. Researchers at the Battelle Institute have summarized the risk factor very well:

> The technical community concerned with risk estimation calculates risks numerically and compares them quantitatively. . . . Moreover, the technical community tends to accept risk estimates at face value.
>
> In contrast, the general public perceives the outcomes of an event to be more important than the probability. This may be due to the fact that the public is familiar with Murphy's Law: If something can go wrong, it will go wrong. Thus, probabilities are often perceived to be less meaningful than outcomes.
>
> However, it is not only the general public that considers outcomes to be more important than probabilities in certain situations. This often occurs in the world of business and industry, and a good example is provided by the nuclear area. Insurance companies would not insure nuclear reactors for unlimited liability because they lacked a data base for actuarial calculations and because one large accident, regardless of how improbable the accident was, could ruin the insurance companies.[1]

LEGITIMACY

The legitimacy of an SSD regime will be raised at the international level because very few nations are developing the capability to dispose of HLW in the sub-seabed, while the consequences of their actions could seriously affect the well-being of a large number of nations. Thus a major issue at the international level is who is to decide whether, when, where, and how to dispose of HLW in the sub-seabed. In addition, how can participation in the decision-making process be assured for the large number of nations who fear that their populations can be adversely affected by the actions of the very small number operating in one of the global commons? Gene Rochlin has posed what he considers the crucial policy decision in the design of an SSD regime: "The fundamental choice to be made is the balance to be struck between the assumption (and sometimes promotion) of equity of treatment for all states and the unilateral or multilateral application of power."[2]

Apart from the question of participation, concern in the international community as a whole will emerge because the global commons are involved. The sites currently being discussed and investigated most intensively for SSD, both in the Pacific and Atlantic Oceans, are all beyond the limits of national jurisdiction as described in the U.N. Convention on the Law of the Sea of 1982.[3] Furthermore, these are areas deemed to represent the common heritage of mankind by the General Assembly in 1970,[4] for which a special regime for the exploration and exploitation of mineral resources has been negotiated in UNCLOS III. In addition, disposal of HLW in the sub-seabed will have to be judged permissible under current international law as stated in the London Convention.

We argue that the international community as a whole has legitimate concern in the SSD option and that a regime wider than the NEA will be required. Thus the last component of the legitimacy issue at the international level which must be raised relates to the general approach to design of an SSD regime, including procedures for review of risk estimates, standards, operations, and the like. Questions such as the following must be answered: What tasks are necessary and desirable at the global, regional, and/or national levels? Why? Relative to the tasks to be performed, are the allocations feasible? If not, are effective adaptations available? (These questions will be discussed in detail below.)

LIABILITY

Liability is a component of the SSD issue because of the large gaps in international law concerning pollution and the damage it causes. As a result, proving damage from radioactive pollution may be very difficult, especially since the effects may be delayed. The problem of proof has recently been summarized by Martine Rémond-Gouilloud as follows:

First, in pollution law a problem of proof often results from the dilution of the effluent in the sea. . . . Yet, in order to obtain compensation, a victim must prove not only his loss but also the origin of the loss; in other words, the source of the pollution and more particularly the chain of causation between the source of pollution and the damage. If he is unable to establish that the loss was caused by a specific agent, he will obtain no compensation. Besides, most legal systems require him to prove that the discharge complained of is a result of a wrongful act or omission ("fault"), in the absence of which the party causing the damage cannot be required to make reparation.[5] *

Clearly any decision on a regime governing SSD will necessarily imply the negotiation of a specific convention on liability. Such negotiations are usually difficult. For instance, the Convention on Liability for Damage Caused by Objects Launched into Outer Space took about ten years to complete. Moreover, there is additional time lost when some signatories are reluctant to see a convention enter into force because they dislike one or another of its provisions. In specialized conventions on liability negotiations in the following areas tend to be very difficult: the definition of damage, the limits of liability, the adequacy of civil versus criminal penalties as effective deterrents, the meshing of different national legal requirements, and the question of compulsory arbitration. Liability negotiation for SSD will amount to a major, time-consuming undertaking in itself.

URGENCY

The urgency of the need for a decision on the SSD option becomes more complex as worldwide nuclear generating capacity increases (see Table 6). Three categories of participants are involved. The first, the nuclear weapons states—i.e., France, the Soviet Union, United Kingdom, United States, China, and perhaps India—all have

---

*In two respects the problem of pollution of the ocean by radioactive wastes is different from other pollution. First, dilution of the effluent is less of a problem given existing capabilities to detect much smaller concentrations of radioactive isotopes than chemical toxins. Second, the discharge of hazardous substances is ipso facto a fault. However, the chain of causation is always very difficult to prove, and in the case of radioactive wastes (as noted) it is complicated by delayed effects.

Table 6

WORLDWIDE OPERATIONAL NUCLEAR GENERATING CAPACITY

Country	Number of Power Reactors Operational	Operational Capacity (GW)	Percent of World-wide Operational Capacity
1. Argentina	1	0.3	< 1%
2. Belgium	3	1.6	1
3. Bulgaria	2	0.8	< 1
4. Canada	10	5	4
5. Czechoslovakia	2	0.5	< 1
6. GDR	6	2	2
7. Finland	3	1.5	1
8. [France][a]	16	8	6
9. [India]	3	0.6	< 1
10. Italy	4	1.4	1
11. Japan	22	0.5	11
12. Netherlands	2	14.5	< 1
13. Pakistan	1	0.1	< 1
14. South Korea	1	0.6	< 1
15. Spain	3	1	< 1
16. Sweden	6	4	3
17. Switzerland	4	2	2
18. Taiwan	2	1	< 1
19. [UK]	33	7	5
20. [USSR]	30	14	11
21. FRG	13	9	7
Total non-U. S.	167	76	41
22. [United States]	74	54	41
Worldwide total	241	131	100

Source: Kidder, Peabody and Co., "Electric Utility Generating Equipment: Status Report on Worldwide Nuclear Reactors" (October 1979); cited in Bupp, pp. 60-61.

[a]Brackets have been placed around nuclear weapons states.

promising domestic terrestrial alternatives. French interest in the SSD option has varied, while the USSR is on record as opposing it in the early stages. As noted, the United States has a small program investigating the feasibility of SSD, and China and India are not known to have made any pronouncements on the issue.

The second category of participants consists of non-nuclear weapons states possessing a commercial and/or research nuclear capability. This category would include seventeen of the twenty-two states shown in Table 6. Of the first two categories of states, the fifteen shown in Table 7 will have the greatest need for disposing of substantial quantities of HLW and LLW. The third category of participants consists of all other states in the international system and a number of intergovernmental organizations (IGOs) and international nongovernmental organizations (INGOs).

Table 7

COUNTRIES WITH GOOD PROSPECTS FOR NUCLEAR
GENERATING CAPACITY GREATER THAN 5 GW AT END OF 1980s

Country	Probable Capacity (GW)
1. [United States] [a]	
2. [France]	50 - 150
3. [USSR]	30 - 40
4. FRG	30 - 40
5. Japan	25 - 35
6. [UK]	10 - 15
7. Canada	10 - 15
8. Sweden	10
9. GDR	5 - 10
10. Spain	5 - 10
11. Taiwan	5 - 7
12. South Korea	4 - 8
13. Switzerland	4 - 7
14. Belgium	4 - 6
15. Brazil	3 - 6

Source: Kidder, Peabody, and Co., "Electric Utility Generating Equipment: Status Report on Worldwide Nuclear Reactors" (October 1979); cited in Bupp, pp. 60-61.

[a] Brackets have been placed around nuclear weapons states.

71

It is clear that the greatest urgency for SSD is felt in the northeast Atlantic region because land-based options may be available to only a few of the participants there and domestic political opposition of a host country would probably make it impossible for one West European state to dispose of HLW within the land territory of another. If an SSD option were determined feasible by the SWG by 1988 or even 1990, one would expect the focus to turn to the sites evaluated as satisfactory in the northeast Atlantic. While the United States is unlikely officially to be a leader in the push toward SSD, one would expect substantial pressure on it to support the SSD alternative because it would solve some pressing problems for U.S. allies.

Pressure on the United States to support an SSD option would increase the tension between the urgency for SSD as seen by a particular region and a likely global preference for greater caution. Regional/global processes would therefore be out of phase, generating further complications for the legitimacy component of the SSD issue, and all players at the global level would be required to take a stand. Early careful and systematic attempts to educate the international community on the SSD option may well mitigate though probably not eliminate this tension.

INTERNATIONAL ANALOGS TO THE SSD ISSUE
AND POTENTIAL LINKS TO RELATED ISSUES

The most important characteristic of the SSD issue as characterized above is that it implies the elimination of undesirable consequences ("bads") rather than the allocation of resources or the distribution of benefits ("private goods"). While this does not mean that the option will generally be seen as providing a collective good, it makes it a potentially easier problem to deal with at the international level, even though the educational task is massive. At the same time, difficulties arise from the fact that while radioactive waste disposal is a major international problem, only a few countries are responsible for creating it. Moreover, both protagonists and antagonists of nuclear power consider effective disposal systems to be inevitably linked to the question of future growth of nuclear power. An additional difficulty with the SSD option is that if feasibility is demonstrated, *initial* urgency may be great only for West European countries. Feasibility will be perceived to have been demonstrated if estimated risks are low, but uncertainty is likely to remain high until sufficient experience with the system has been gained. Yet public concern for the effects of accidents, even if risks are low, will be a major problem.

There are not many analogs to the SSD issue in the international system, so inferences about the dynamics one could expect are limited. However, it will be useful to compare the SSD issue with the following: (1) negotiating resources regimes for the Antarctic; (2) the Nuclear Non-Proliferation Treaty (NPT); (3) UNCLOS III, and (4) global environmental protection issues.

RESOURCES REGIMES FOR THE ANTARCTIC

As with SSD, the aim of the original international regime for the Antarctic, known as the Antarctic Treaty and negotiated in 1959, was principally to eliminate "bads"—specifically to avoid disputes over conflicting territorial claims and to prevent the extension of superpower conflict to the Antarctic, which was only marginally useful militarily but very important scientifically. As with SSD, the disposition of the issue was of primary importance to only a few states, and the regime worked well for about a decade. In the 1970s, however, the stability of the regime was threatened by a combination of factors: (a) changes in the law of the sea, particularly in the notion of exclusive economic zones (EEZs) of 200 miles; a potential conflict brewed between claimant states with their presumed rights in the EEZs and others who felt that the common heritage principle should be applied to the Antarctic; (b) shifts in fishing effort to species of the Southern Ocean, especially krill, by large distant-water fleets suffering the effects of exclusion by coastal states from areas in which they had previously been accustomed to fish;* (c) changes in the law of the sea defining the limits of the continental shelf, and the rapidly increasing salience of access to oil after the oil embargo in 1973 by the Oil Producing and Exporting Countries (OPEC).

---

*The shift in fishing effort to krill affects whales and other marine mammals, for whom krill are the main food source.

It is interesting to note two consequences of these changes. First, a continued suspension of territorial claims in the area required the maintenance of a closed regime in the eyes of most claimant states—especially those in Latin America. Moreover, the Latin Americans were highly influential members of the Group of 77 in UNCLOS III, making it easier to avoid the application of the common heritage principle to the Antarctic when the issue was pushed by Sri Lanka, who was host to a meeting of the Non-Aligned Nations in 1976. Second, the continuation of a closed regime accommodating the interests of the claimant states made it easier for states with distant-water fishing fleets operating in the Southern Ocean and either signatories of or parties to the Antarctic Treaty to insist on a loose regime governing the exploitation of living resources in the Antarctic, even though uncertainty about the status and dynamics of krill remains high.

In the case of the SSD option, a closed regime is likely to face a much greater challenge from the international community than did the Antarctic Treaty as a result of concern with the potential effects of accidents combined with a growing and sometimes strident anti-nuclear movement in Western Europe, North America, Japan, and the west-central and southwest Pacific. This movement is fast becoming transnational and is a significant constraint for governments on issues of nuclear power and the disposal of radioactive waste.

THE NUCLEAR NON-PROLIFERATION TREATY

The NPT is another instance of a regime that was aimed primarily at the elimination of "bads," but from the beginning the NPT carried a far larger burden than the Antarctic Treaty. For one thing, the global salience of the issue was high, and it contained a significant dimension of North/South conflict, even though in this case the Soviet Union was perceived to be a member of the North. By implication, another gap in the treaty became apparent in the emergence of a significant conflict internal to the North/West group (primarily members of the OECD) over the competition in the sale of reactors and other portions of the nuclear fuel cycle to third parties.

The North/South conflict centered around the lack of credibility of security guarantees in the treaty, the unwillingness of the super-powers (according to others) to take meaningful steps toward general and complete nuclear disarmament, and the lack of incentives to join the NPT, which called for renunciation of the right to own nuclear weapons, compared to the continued rewards of not becoming a signatory. Thus while the NPT is a global regime aimed at the elimination of "bads," it is of questionable effectiveness because it raises significant questions about the distribution of benefits and because the penalties of membership are increasingly considered unfair, both in the face of superpower intransigence and in the lack of credible controls placed on non-signatories. Unlike the Antarctic Treaty, the nature of the regime itself has become a significant source of conflict.

UNCLOS III

The new ocean regime negotiated within UNCLOS III is multi-dimensional, but at its heart lies a compromise between the interests of coastal states for extending sovereignty and exclusive jurisdiction over resources and activities of economic and other significance in the EEZ and the interests of the international community in safeguarding free navigation. The convention has also elaborated a unique regime to govern the exploration and exploitation of mineral resources on the seabed beyond national jurisdiction, and it has elaborated a set of standards relating to the protection and preservation of the marine environment. While land-based sources of marine pollution are included, the remedies prescribed are unlikely to be effective. The most significant parts of the convention in this context have to do with coastal states' rights to regulate pollution in "special" and "ice-covered" areas and enforcement actions against ship-generated pollution by coastal, port, and straits states.

The SSD issue is a natural subset of law-of-the-sea issues in two respects: (1) it is related to the international seabed regime and the prerogatives of the International Seabed Authority; and (2) SSD must adhere to the provisions relating to the protection and preserva-tion of the marine environment, including by extension the question of whether the London Convention applies to SSD. In addition, the future of the SSD option will be affected by two ancillary issues raised by UNCLOS III. The first is the future of the ocean enclosure movement represented by the EEZs and the long-term stability thereof in the face of continued technological advance. The second concerns the nature of the entry-into-force of the Law of the Sea Treaty and the fact that some states may neither sign nor ratify.

The ocean enclosure movement will be a long-term problem only if the treaty does not come into force. If it fails, the EEZ movement will not automatically come to rest at 200 miles. Even if the treaty comes into force, coastal states will still try to extend their jurisdiction beyond the EEZ, but it will be easier to contest their claims with a treaty than without. If the treaty, which was completed in December 1982, is signed and ratified by the member states of the OECD—including the United States—it will be easier to work for an accommodation between the SSD regime and the International Seabed Authority because (as noted) the Group of 77 does not necessarily have a single point of view on SSD. Since the issue of the jurisdiction of the International Seabed Authority will not have to be raised, the SSD issue will not need to be cast in terms of the North/South conflict. Conversely, if the treaty is not signed and ratified by at least some of the OECD members—including the United States—it will tend to complicate the SSD issue because a wider conflict will ensue relating to the jurisdiction of the International Seabed Authority—especially if the non-signatories of the OECD proceed to implement a "Reciprocating States" regime relative to the exploration and exploitation of the minerals of the international seabed area. Even if some member states of the OECD reject the treaty and implement a Reciprocating States regime, they could presumably proceed, or attempt to proceed, with developing the SSD option for the north Atlantic alone, but they would be subject to enormous pressure diplomatically and domestically and would possibly even be challenged through the courts. Clearly the consequences of an accident occurring in such an atmosphere of conflict would be magnified.

GLOBAL ENVIRONMENTAL PROTECTION ISSUES

It appears that the SSD issue is most similar to certain global environmental issues like toxic waste disposal, limitation of carbon dioxide in the atmosphere, reduction of acid rain, and depletion of the ozone layer. However, no large-scale regime has yet been designed or negotiated for these issues. It therefore appears that for issues analogous to that of SSD, conflict over a possible regime is highest where substantive links exist to resource allocation or distribution of benefits and/or where principles of coastal state jurisdiction are involved. Unless an SSD regime becomes a point of contention in itself, the major questions will revolve around regime design, standards, liability, and enforcement. If a conflict with the Law of the Sea Treaty can be avoided, the major sources for potential regime conflict will lie in the application of the London Convention and the growing anti-nuclear movement. The issues concerning the application of the London Convention have already been analyzed in detail by others.[6] Our concern here is to determine whether SSD constitutes dumping and is therefore prohibited by the London Convention. By extension, this also involves the Convention on the Law of the Sea of 1982 because the definition of dumping contained therein (Article 1) is only a slightly modified version of that contained in the London Convention.

Currently the best assessment of whether SSD constitutes dumping is *probably not*, though some ambiguity remains.[7] The United States has not expressed an official position as yet, but some U.S. officials feel that neither the London Convention nor the Law of the Sea Treaty applies to SSD, assuming that no harmful radioactivity is released after emplacement.[8] This implies that emplacement itself cannot be construed as pollution or dumping and that the test to be applied after emplacement of the canisters is twofold: (a) There must be no release of radioactivity greater than background levels;* and (b) No adverse effects to the marine environ-

---

*Formerly a unit of radioactivity was measured in curies (Ci). Now the new unit of measurement is called a becquerel (Bq), which corresponds to one nuclear disintegration per second ($1Ci = 3.7 \times 10^{10}$ Bq). The oceans contain a substantial amount of naturally occurring (i.e., background) radioactivity whose origin is either terrigenous or cosmic. More than 90 percent of naturally occurring radioactivity in the oceans comes from Potassium 40 (^{40}K) in an amount estimated to be 1.5×10^{22} Bq. By contrast, man's contribution is estimated to amount to 0.7 percent of natural radioactivity (see P.K. Park et al., "Radioactive Wastes and the Ocean: An Overview," in *Radioactive Wastes and the Ocean*, ed. P.K. Park et al. (New York: John Wiley and Sons, 1983), vol. 3, pp. 22-25.

ment result from such releases. In addition, at least informally in U.S. government circles, there seems to be the position that if canisters can be retrieved, disposal ipso facto cannot constitute dumping. Indeed during the Carter administration, the view was often expressed that unless canisters could be retrieved, the SSD option would be ruled out. Alternatively other officials have argued that as the London Convention is currently worded, SSD falls within the definition of dumping.

Two points need to be made here. First, the ambiguity is sufficiently important that it requires some resolution, perhaps in the form of an amendment to the London Convention. Second, the notion that SSD does not constitute dumping if canisters can be retrieved needs further clarification. For one thing, as John Norton Moore points out, it is not in the U.S. interest to argue for a policy which will permit *unilateral* resort to SSD beyond zones of national jurisdiction because there will be no guarantee that requisite safety and environmental protection standards will be met.[9] For another, such unilateral resort to SSD will not be considered legitimate by most of the rest of the world; if it is pursued internationally, it will run a substantial risk of triggering a conflict on principle which in turn could unite the transnational environmental protection and anti-nuclear movements.

Since by extension the question of dumping also involves the Convention on the Law of the Sea, we need to ascertain whether there is anything in the treaty which can be construed as prohibiting the disposal of HLW in the sub-seabed. There is not. The jurisdiction of the International Seabed Authority is confined to the exploration and exploitation of the resources of the international seabed area. "Resources" in this context are defined as "all solid, liquid or gaseous mineral resources in situ in the area at or beneath the sea-bed, including polymetallic nodules" (Art. 133(a)). However, the International Seabed Authority is explicitly given jurisdiction over protection of the marine environment (Art. 145), although this jurisdiction is limited because it encompasses only activities relating to exploration and exploitation of resources as defined above. At the same time, within the range of its jurisdiction, its rule-making powers are defined broadly (Art. 145(a)), including the competence to

adopt appropriate rules, regulations and procedures for *inter alia:*

(a) the prevention, reduction and control of pollution and other hazards to the marine environment, including the coastline, and of interference with the ecological balance of the marine environment, particular attention being paid to the need for protection from harmful effects of such activities as drilling, dredging, excavation, disposal of waste, construction and operation or maintenance of installations, pipelines and other devices related to such activities (Art. 145).

It is clear therefore that while the International Seabed Authority does not have the competence to be a central player in the SSD option, there will at least have to be consultation with it on risks to the marine environment beyond national jurisdiction. Such consultation can be deemed required by Article 147 (3): "Other activities in the marine environment shall be conducted with reasonable regard for activities in the Area."

Apart from concerns for protection of the marine environment, is SSD likely to trigger serious conflicts with other foreseeable uses of the international seabed area? If it is, the International Seabed Authority will have a substantial role, as stipulated not only by Article 147 (3), but also by Article 137 (1):

No State shall claim or exercise sovereignty or sovereign rights over any part of the Area or its resources, nor shall any State or natural or juridical person appropriate any part thereof. No such claim or exercise of sovereignty or sovereign rights, nor such appropriation shall be recognized.

The area in which canisters are emplaced will ipso facto constitute appropriation because it will exclude others from using the area for other purposes. But such appropriation is not necessarily a problem, both because the total area to be used for SSD is small and because site selection criteria will rule out locations which are of major potential economic significance. The SDP has made a preliminary assessment of this possible conflict and drawn the following conclusions:

1) In the near future there is no conflict between deep ocean mining and subseabed disposal, since they are considering different parts of the oceans.

2) If deep ocean mining were to expand to all the nodule covered areas of the ocean, the area occupied by a repository would amount to a trivial loss in mining operation area, about 0.0005%.

3) Should for any reason a nodule mining operation be conducted in the area of a repository, it would not impair the containment capabilities of the repository.[10]

Perhaps, as we have noted, the critical issue-linkage problem for the SSD option will emerge out of the multidimensional anti-nuclear movement. Because the anti-nuclear issue is very complex and at times emotion-laden, it will be very difficult to deal with. Its components are the following:

1) A growing public opposition in Western Europe, North America, and Japan to the size and potential destructiveness of the nuclear forces maintained by the superpowers and some of their allies.

2) A potential linkage in the United States between the desire of the Reagan administration to solve the nuclear waste disposal problem (as the sine qua non of an effective strategy to promote the commercial use of nuclear energy) and a growing public resistance to the further growth of nuclear energy by publics in the United States and Western Europe.

3) A potential linkage between the security interests of the United States and Western Europe in avoiding too great an energy dependence on the USSR and concern to avoid escalating nuclear proliferation as a result of promoting nuclear sales abroad and pursuing a policy of commercial reprocessing of spent fuel.

4) Sustained high-level political opposition in Japan to the Japanese government's stated intention to dump LLW in the ocean approximately 600 miles north of Guam. This opposition merges the anti-nuclear movement in Japan with the large and powerful constituency of Japanese fishermen, who fear adverse effects from dumping on the fish stocks which constitute their livelihood.

5) A widespread and intense opposition among governments and peoples of west-central and south Pacific island states to the

storage and/or disposal of all forms of nuclear waste in the Pacific. This opposition appears to be inspired by a concern for adverse effects on fish stocks as well as highly emotional resentment of past and present U.S. and French nuclear testing in the Pacific.

These components are an extremely potent potential mix which is not now lined up specifically against any one nuclear waste disposal option in a comprehensively integrated fashion.[11] For any option which is beyond national jurisdiction, the emerging transnational links among these components will be a very important constraint. For this reason, we argue that those who wish to pursue an SSD option pay attention to the need for global legitimacy and invest in educational programs, as well as satisfy the international community on issues related to reliability and liability. More specifically, it is important to recognize that this mix of anti-nuclear components results in an ineluctable connection between the issues of dumping LLW in the ocean and the SSD option for HLW, both in the Pacific and in the Atlantic. Care should therefore be taken that a comprehensive evaluation of the feasibility of SSD not be stopped prematurely as a result of intense political conflict generated by the pursuit of the dumping option for LLW.

LIKELY MAJOR PROBLEMS OF NEGOTIATING AN SSD REGIME

SITE SELECTION AND METHOD OF EMPLACEMENT

The location of disposal sites will present problems of political acceptability that will have to be dealt with, at least in part, by negotiation. It is currently thought that the most promising site in the world is in the northwest Pacific, but this site poses the greatest problems of political acceptability. In addition, owing to the sensitivity of the issue, states bordering the north Atlantic would probably not be allowed to dispose of HLW in the north Pacific, even if transportation costs were tolerable. For reasons that appear related primarily to the intensity of domestic opposition, Japan in March 1982 at least temporarily decided that disposal of LLW in a site south of the northwest Pacific potential SSD site was not politically feasible (see point (4) on p. 80 above).

It is important to realize that site selection decisions present problems of volatility because both knowledge and political factors are highly dynamic. For instance, in the early stages of the SWG, only three Mid-Plate/Gyre (MPG) potential sites were identified. MPG-I was located in the north-central Pacific about 600 miles north of Hawaii; MPG-II was a large area in the northwest Pacific; and MPG-III was a large area in the central and west-central north Atlantic. As additional work was done, more potential sites were identified in both oceans, and thus new selection decisions became necessary. (The sites are listed in Table 8.)

From the research done to date, firm conclusions cannot yet be drawn regarding the most likely sites for SSD. However, at a meeting of the SWG in March 1982, the Site Selection Task Group agreed that on the basis of what was then known, the type of site likely to prove most acceptable was the distal abyssal plain. At the time, the task group had in mind a model of shallow emplacement and had not yet studied in detail models of deep emplacement which would change evaluations of site potential. Under these circumstances, the sites which looked most promising (in order of acceptability) were MPG-II in the northwest Pacific, the Great Meteor East, and the Nares Abyssal Plain in the west-central Atlantic.

Japan's decision to postpone the dumping of LLW in the northwest Pacific seems to underscore our asssumption that the first disposal site for SSD is likely to be in the Atlantic, where the need and urgency of countries is greater than in the Pacific. But if one takes the long view, clearly domestic policy relative to the SSD option in the Pacific will be considerably affected by the lessons learned in the Atlantic. In this light, if Japan's decision is sustained, it may serve to inhibit a potential connection between the dumping of LLW and SSD in the Pacific. If so, the northwest Pacific SSD site may well be used in future.

Choice of sites in the Atlantic appears to present fewer political problems, not only because the need is greater, but also because significant internal opposition in the major states concerned has not yet crystallized to the same degree as in the Pacific. However, it is instructive to note that opposition to MPG-II in the Pacific did not originate as a result of Japan's interest in the SSD option (which is low), but in its stated intentions to dump LLW in the northwest Pacific. If the United Kingdom, the Netherlands, Belgium, and

Table 8

POTENTIAL SITES FOR SUB-SEABED DISPOSAL OF HLW

Potential Site	Region	Specific Location
Pacific Ocean:		
*PAC 1 (MPG-II)[a]	Northwest Pacific	20-50°N; 143-180°E
PAC 2 (MPG-I)	North-central Pacific	20-35°N; 140-160°W
PAC 3	Northeast Pacific	20-35°N; 115-130°W
*Nares Abyssal Plain	Northwest Atlantic	22.5-23.6°N; 63.0-64.3°W
*Great Meteor East	Northeast Atlantic	31.2-31.8°N; 24.2-25.0°W
Cape Verde (CV) Plateau:		
CV_1	Northeast Atlantic	~25°N; ~25°W
CV_2	Northeast Atlantic	19.2-19.5°N; 29.8-30.0°W
Great Meteor West	North Atlantic	30-31°N; 29-30°W
Northern Bermuda Rise	Northwest Atlantic	32-35°N; 56-64°W
South Sohm Abyssal Plain	North Atlantic	32.0-32.5°N; 55-56°W
South Hatteras Abyssal Plain	Northwest Atlantic	~26°N; ~70°W
CV_3	Northeast Atlantic	23-26°N; 24-28°W
Gambia Abyssal Plain	Northeast Atlantic	~11°N; ~28°W
Southern Bermuda Rise	Northwest Atlantic	29-32°N; 64-69°W
Mad Cap	Northeast Atlantic	28.5-30°N; 24-25°W
King's Trough Flank	Northeast Atlantic	41.0-42.5°N; 21-23°W
Iberia Abyssal Plain	Northeast Atlantic	~40°N; ~140°W

Sources: E. Laine et al., "Site Qualification Studies," in Sandia, *Annual Report . . . 1980;* Woods Hole Oceanographic Institution, *Minutes of the Site Assessment Committee Meeting,* 16 July 1981; and K. Hinga, personal communication, 16 June 1982.

[a]Asterisks indicate most promising sites. MPG is identified on p. 82 above.

Switzerland insist on dumping LLW in the north Atlantic, the future of SSD could be even more doubtful. The point here is not whether such dumping presents a significant radiological hazard to human beings and marine ecosystems; rather the point is that LLW dumping has proved to be a powerful emotional trigger for catalyzing politically potent opposition from the different components of the anti-nuclear movement, and to that extent, it is a hindrance to further explorations of the feasibility of the SSD option.

Even though site selection for SSD may not be as difficult in the north Atlantic as it is in the northwest Pacific, variations in the location could present thorny negotiation problems nonetheless. For one thing, if the site is anywhere in the north Atlantic, the Scandinavian countries will have to be convinced that the SSD option does not constitute dumping. For another, if a site were particularly close to Portugal and Spain, it would present a more difficult problem than one nearer the north-central Atlantic. (The Spanish government and to a lesser extent the Portuguese government have adopted very strong positions against the dumping and disposal of all radioactive waste in the ocean.)

While site selection may make negotiations more or less difficult, it does not change our basic argument regarding the design and creation of a widely acceptable international regime. Whichever site is chosen first will have significant implications for all other potential sites, and the way it is handled will help to determine whether an SSD regime becomes a highly contentious issue in the international system. We argue that a high level of contention is not foreordained and that care must be taken to avoid it.

There are currently three options being investigated for the method of emplacement: shallow emplacement (free fall), shallow emplacement with dynamic penetrator, and deep drilling. If in the foreseeable future the deep drilling option is chosen, it will have two consequences. First, some sites currently thought not to be promising are likely to become so, thereby presenting different site negotiation problems. Second, effective measures will need to be found for insulating the International Program of Ocean Drilling (IPOD), which does marine scientific research, from the operational and regulatory needs of SSD. The negotiation of measures will be particularly sensitive if a vessel is chartered for deep drilling and is at the same time utilized by IPOD.

REGIME DESIGN AND CREATION

Major problems of regime design and creation will be discussed in terms of four questions:

(1) What do we know about the dynamics of regime creation and maintenance in the international system generally?

(2) What lessons can be inferred from our discussion on the design and creation of an SSD regime?

(3) How should criteria for SSD regime design be derived?

(4) What allocation of responsibilities and resources is likely to yield the most effective result at the global level?[12]

(1) *Regime creation and maintenance.* International regimes have been succinctly defined by Ernst Haas:

Regimes are norms, procedures, and rules agreed to in order to regulate an issue area. Norms tell us *why* states collaborate; rules tell us *what*, substantively speaking, the collaboration is about; procedures answer the question of *how* the collaboration is to be carried out. Procedures, therefore, involve the choice of whether specific administrative arrangements should be set up to regulate the issue area. Administration involves organization.[13]

Haas distinguishes among four types of procedures: those that require only a common framework for the pooling of capabilities and exchange of information; those that require a joint facility which standardizes behavior on the basis of common routines; those that require a common policy which adjusts behavior to the planned needs of a collectivity; and those that require a single policy which is centrally developed at the international level.[14] These types represent an ascending order of complexity; the last two imply fairly elaborate organizational arrangements.

In order to create regimes, significant political mobility is required to establish three types of coalitions which interact: transgovernmental, transnational, and intergovernmental. A transgovernmental coalition exists when functionally similar government agencies collaborate in spite of or in opposition to a policy enunciated by the Ministry of Foreign Affairs of one or more of the governments concerned.[15] A transnational coalition includes at least one member who is not a government or a party thereof.[16] Intergovernmental organizations are both actors and arenas in such coalition politics. They help set agendas, act as catalysts for coalition formation, and are fora in which attempts can be made at a broader linkage of issues by states lacking capabilities on the subject being negotiated.[17] The choice of organizational fora for dealing with issues in a regime is therefore very important. Furthermore, it is necessary to know who has power over outcomes in each forum. Power over outcomes on issue X in organization Y is not necessarily transferable to issue B in organization A.[18] One needs to know whether such power is determined solely by the mobilization of winning coalitions within a particular forum or whether the capacity to affect choices and behavior outside the forum is more important.[19]

With respect to issues on an international regime agenda, Haas makes the significant point that the narrower the scope of issues to be negotiated, the higher the degree of certainty about efficient solutions because benefits and costs can be calculated fairly reliably if the knowledge base is firm.[20] In contrast, if the knowledge base is not firm and uncertainty is high, even experts will be tempted to expand the scope of issues as a check on the efficacy of broader types of solutions.[21] High uncertainty will therefore tend to generate a propensity to link issues, making it harder to arrive at effective solutions because the calculation of benefits and costs will be complicated.

Attempts at issue-linkage usually do not succeed if states with strong stakes in an existing distribution of benefits and costs and the capability to control that distribution prefer things as they are.[22] For this reason, coalitions of states possessing the greatest capabilities favor issue-specific negotiations. Conversely, fragmented issue-linkage is favored by challengers who seek to obtain bargaining leverage and/or maintain the cohesion of their coalition.[23]* Occasionally there occurs what Haas calls substantive issue linkage — i.e., consensual knowledge linked to agreed social goals. But it is important to realize that this occurs "only when the possibility of joint gains from the collaboration exists and is recognized."[24]

(2) *Lessons for designing and creating an SSD regime.* From the discussion above at least two lessons can be inferred. First, uncertainty over the SSD issue is high and likely to remain so until experience with SSD accumulates. For this reason, if the option is

---

*Haas explains that fragmented issue-linkage occurs when a "coalition is held together by a commitment to some overriding goal, even though the partners disagree with respect to the knowledge necessary to attain it" (Haas, p. 372).

deemed technically feasible and some members of the OECD wish to adopt it as an alternative or even an "only way to go," we envisage that they will favor issue-specific negotiations in a forum they dominate, while states lacking a capability to evaluate and eventually to utilize SSD may wish to broaden the scope of issues. In so doing, they may wish to pursue issues which we have described as potentially linked to SSD.

Second, since the knowledge base on SSD is not yet firm and will not be at the regime-creation stage, experts within the OECD member states most concerned may also be disposed to seek a broader agenda by linking issues. If this occurs, considerable uncertainty will be generated, and the SSD issue may come to be regarded politically as a Pandora's box. However, we must acknowledge that if a large-scale educational program is successful in demonstrating that significant joint gains can come from an SSD alternative, regime creation will be easier—although no matter what the circumstances, the creation of an SSD regime will be a major undertaking.

(3) *Criteria for SSD regime design.* We recommend that the design criteria be derived from the structure of the SSD issue as analyzed in the first section of this chapter, taking into account the participants, their interests, capabilities, and priorities, and the SSD technology with its attendant uncertainties.

As we have noted, the SSD issue has five components: feasibility, reliability, legitimacy, liability, and urgency. Potential links to other issues will affect these components in different ways. For instance, links to the ocean dumping of LLW raise questions about feasibility and legitimacy; links to UNCLOS III raise questions about legitimacy and reliability; and links to the anti-nuclear issue raise questions about feasibility, reliability, and legitimacy. With regard to the players, we emphasize that among the nuclear weapons states—especially Western Europe, the United States, the Soviet Union, and Japan—interests, capabilities, priorities, and perceptions of urgency differ.

SSD technology poses special problems. Some aspects of the technology and knowledge base are new, and there has been a rapid rate of advance in knowledge and capability. As knowledge increases, assessments of feasibility will change. Since uncertainty is high and the rate of change in the knowledge base is also high, the regime should be designed as a controlled experiment within limits to be defined by the legitimate international authority. As the debate over the International Seabed Authority demonstrates quite clearly, the creation of a flexible international regime requires trust. If trust is absent and each party to a regime fears it is vulnerable to the others, the regime will be rigid and unable rapidly to adjust to changing conditions.

We consider that the four most important criteria for system design are feasibility, reliability, legitimacy, and liability. For countries with a nuclear capability feasibility will be decided primarily at the national level, but the decision process will not necessarily be the same for each. In some cases feasibility may be judged by giving greater weight to technical criteria; in others political criteria may predominate. For countries lacking a nuclear capability, determination of feasibility will have to be made primarily through participation at the international level. Urgency will also be decided at the national level, and we have argued that there may be tension between the pace of global and regional approaches to SSD.

Because global and regional processes may be out of phase, different regime designs will be necessary and desirable at the global versus the regional level. It seems to us that at the global level the most appropriate arrangement would be (in Haas's terminology) a common framework in which the decision process would do the following:

1) Review assessments about the technical feasibility of the SSD option;
2) Decide on the basis of performance characteristics whether SSD is dumping or decide to change the definition of dumping;
3) Set standards governing the design and operation of system components to meet the tests of safe disposal and protection of the marine environment;
4) Maintain a system of inspection and routine checks;
5) Settle the issue of liability.

At the regional level a joint facility already exists in the NEA, but a common policy is needed to treat with the operating criteria—consistent with global standards—for choice of site, design of vessel, transport to site, design of waste form and canister, method of emplacement, and system of monitoring. We stress that no disposal of

HLW in the sub-seabed beyond national jurisdiction should be allowed unless it is the result of a linked global/regional regime.

We think that a global/regional division of tasks is feasible. The global level does not have and is not likely ever to have operating capabilities. Since resources are not to be allocated or benefits distributed, it is possible to avoid conflict over the regime if the four design criteria specified above are employed. The salience of the issue will vary with the location of disposal sites. For political reasons it would be advisable to begin SSD in the Atlantic rather than the Pacific. Then if the system is widely perceived to be successful, it is likely to affect choices for site location in the Pacific.

The operational capabilities are clearly concentrated in the OECD. At the same time, the attitude of the USSR may well be important vis-à-vis the rest of the world. As a result, the participation of both OECD members and the USSR on at least the global level will be desirable. Moreover, concerted attempts to inform the rest of the world at the global level are necessary. Some developing countries will have a significant long-term interest in the demonstrated feasibility (or lack thereof) of the system.

(4) *Responsibilities and resources for SSD at the global level.* The allocation of responsibilities and resources poses a more difficult problem than appears at the surface because IGOs are not isolated units but are always parts of networks which include different national constituencies, other IGOs, and INGOs. At the global level interorganizational conflicts over program priorities and possibilities for task expansion are endemic because the funds to support international collaboration are small, sunk costs are great, and executive heads perceive themselves as being caught in a zero sum game. Such conflicts are usually refracted to national constituencies and back to international organizations. Sometimes these conflicts encompass competing coalitions within or across entire networks, making it very difficult to proceed in a straightforward fashion. Moreover, each organization has its own governing bodies. Decisions of one organization affecting the work of a target organization are not consequential unless the national constituencies of the former have already received the support of the national constituencies of the latter or (if they have not) unless the decisions have already been negotiated with and accepted by the target organization. If either or both of these routes

is/are not pursued, conflict ensues and often spreads to conflict between national constituencies which is time-consuming and costly.

In the UN system only the General Assembly has the authority to allocate tasks and programs centrally across other organizations in the system, but the Assembly knows full well how difficult it is to do so. Indeed the last time the Assembly engaged successfully in such central allocation was in 1961-62 concerning the design of a regime for the exploration and exploitation of outer space; its efforts came because of a strong initiative by the Kennedy administration.[25] Despite the Assembly's authority, the allocations had to be negotiated with the organizations and their national constituencies beforehand, and significant conflict still occurred.[26] In consequence, every time an article in the Convention on the Law of the Sea of 1982 touches upon the role of an international organization, the phrase "competent international organization" is used, thereby deftly sidestepping the issue of central allocation to a specific organization.[27]

The potential players in an SSD regime would include the following IGOs:

1) The signatories of the London Convention;
2) The International Atomic Energy Agency (IAEA);
3) The Intergovernmental Oceanographic Commission (IOC);
4) The United Nations Environment Programme (UNEP);
5) The interorganizational Group of Experts on the Scientific Aspects of Marine Pollution (GESAMP);
6) The International Maritime Organization (IMO);
7) The World Health Organization (WHO);
8) The International Seabed Authority—if and when the Convention on the Law of the Sea enters into force; and
9) The NEA.

Two INGOs would be players:

1) The Scientific Committee on Oceanic Research (SCOR);
2) The Scientific Committee on Problems of the Environment (SCOPE).

Both are components of the International Council of Scientific Unions (ICSU).

The signatories of the London Convention will be critical in the decision on whether SSD is or is not dumping. Similarly, only IMO has the competence to prescribe standards for vessel design, construction, equipment, and manning, and the relevant portions of the Convention on the Law of the Sea confirm IMO's authority. As noted above, liability will have to be dealt with as a separate issue in the form of a convention. This leaves standard-setting relative to system components and inspection as the areas in which it may be difficult to reach international agreement. The argument has been made that the IAEA may be the central unit in the SSD regime since the tasks that must be performed in SSD are largely already being performed by the IAEA as a result of adaptations imposed by the NPT.[28]

The logic of the division of responsibilities outlined above is impeccable, but by itself it will not suffice because organizations without assignments in the potential network will be maneuvering for roles and their executive heads will attempt to mobilize their national constituencies in their favor. It is therefore among the national delegations to IGOs that the allocation of tasks will have to be hammered out. However, in most cases such delegations are coalitions themselves and not monoliths. This means that solutions must be negotiated internally. Moreover, international secretariats have access to the national constituencies on national delegations (as they often do in transnational organizations as well). Further complications may arise if particular government agencies do not like the effects of national decisions and seek to have them defeated in international fora by mobilizing transgovernmental coalitions of like interests.

If the IAEA is indeed to be the center of the SSD regime as a result of its operational capabilities, the advisory relationship of the remaining organizations to the IAEA will have to be spelled out. The potential network that has been identified here is not very large as networks go. It corresponds in size to the network for international marine science led by the IOC and therefore would be smaller than the networks for both fisheries and international shipping.[29]

EPILOGUE

Since this study was written in July 1982, the SWG has created a Legal and Institutional Task Group (LITG) for assessing the legal, political, and institutional feasibility of the SSD option.* In its consideration of the issue, the group has come to the following preliminary conclusions:

1) A sharp distinction should be drawn between the research/experimental and operational phases of SSD because very different conditions pertain to each phase and different legal regimes apply.

2) The London Convention does not apply to the research/experimental phase.

3) The London Convention *as currently drafted* should not apply to the operational phase of an SSD program.

4) The NEA and SWG will determine only the concept feasibility of the SSD option. A decision to implement a program will have to be made by the international community as a whole under procedures yet to be determined. Even if the concept proves to be feasible in the late 1980s or early 1990s, the international community will have up to two decades in which to decide whether—and if so, how—SSD should be implemented.

5) The Convention on the Law of the Sea of 1982 does not prohibit the disposal of radioactive wastes in the seabed if it is conducted in an environmentally safe manner. However, if and when the treaty comes into force, the International Seabed Authority will at least have to be consulted.

6) General principles of international law, apart from treaty law, do not prohibit either the research/experimental or operational

---

*As one of the two U.S. representatives to the LITG, Edward Miles was elected chairman.

phases of an SSD program. However, two general principles would apply to an operational phase:
   a) States have a duty to avoid causing damage to the international marine environment so as not to infringe on the rights of other states.
   b) States are responsible and liable for any damage so caused.
7) There are three options for establishing the operational phase of an SSD regime:
   a) Amend the London Convention explicitly to cover SSD.
   b) Draft and negotiate a new global treaty.
   c) Amend the Convention on the Law of the Sea of 1982 to apply to SSD.
8) Whichever option noted in (7) above is chosen, an SSD regime must provide for strong safety and performance standards and create an effective review and monitoring system.[1]

As we have predicted, the future of the SSD option will be shaped—even in the research/experimental phase—by two sets of external events: The first relates to the growing controversy—and indeed within and between certain governments, confrontation—on the issue of LLW dumping. The second relates to the rejection of the Convention on the Law of the Sea by the United States.

At the Seventh Consultative Meeting of the Contracting Parties to the London Convention (14-18 February 1983) a resolution was adopted calling for suspension of ocean dumping of radioactive waste until proposals to amend the technical annexes to the convention could be considered in light of the opinion of a scientific ad hoc group to be set up under IAEA auspices.[2] The Netherlands and the United Kingdom argued that this moratorium was not binding because its adoption did not follow agreed-upon procedures and that they intended to proceed as planned with dumping in the summer of 1983. Furthermore, they warned that if actions continued to be taken in violation of agreed-upon procedures, the fabric of the London Convention could be damaged.

During the summer of 1983 the focus of a transnational anti-dumping coalition was on national decisions in the Netherlands and the United Kingdom. Conflict between the coalition and the two governments was so intense that the Dutch government appeared to back down completely and agreed to search for terrestrial disposal possibilities. The United Kingdom did not back down, but it was prevented (perhaps temporarily) from dumping by a strike of transport and dockyard workers.

These events clearly demonstrate that transnational (and transgovernmental) anti-dumping coalitions can be effective players at both the global and national levels. Furthermore, as some governments insist on dumping LLW in the ocean, the conflict will become more heated and strengthen the coalitions. If the conflict spreads, the future of even the research/experimental phase of the SSD option will be in doubt.

There are governments and nongovernmental organizations within the anti-dumping coalition who would like to get a formal decision from the contracting parties to the London Convention that SSD is covered by Article III of the convention and is therefore prohibited. At the end of the Seventh Consultative Meeting Norway, Spain, and the Federal Republic of Germany introduced a resolution which referred to the need for "clarifying if, and eventually under what circumstances, sea-bed disposal of high-level radioactive waste would be contrary to the provisions of the Convention."[3] Moreover, it called for a legal and technical expert group to meet intersessionally and report to the Eighth Consultative Meeting in February 1984. (The meeting of experts was later scheduled for 12-14 December 1983.) The Group of Experts was unable to achieve consensus on a draft resolution and could agree only that "the Consultative Meeting of Contracting Parties to the London Dumping Convention is the appropriate international forum to address the question of the disposal of high-level radioactive wastes into the seabed, including the question of the compatibility of this type of disposal with the provisions of the London Dumping Convention."[4]

At the Eighth Consultative Meeting (20-24 February 1984) the contracting parties to the London Convention could not agree on the question of the applicability of the convention to SSD of HLW. However, they were able to agree on the statement of the Group of Experts, as well as on the following:

No such disposal should take place unless and until it is proved to be technically feasible and environmentally acceptable, including a determination that such waste can be effectively isolated from

94

the marine environment and a regulatory mechanism is elaborated under the London Dumping Convention.[5]

The issue was then referred to the Ninth Consultative Meeting scheduled for September 1985.

No overt links have yet been made between the SSD issue and U.S. rejection of the Convention on the Law of the Sea of 1982. However, these will be unavoidable if the SSD option survives into the period when the treaty enters into force and the International Seabed Authority is created.

NOTES

Chapter 1: Introduction

1. For general literature on HLW disposal, see the following: Arthur S. Kubo and David J. Rose, "Disposal of Nuclear Wastes," *Science* 182, 4118 (21 December 1973):1205-12; Gene I. Rochlin: *Plutonium, Power and Politics* (Berkeley: University of California Press, 1979), and "Nuclear Waste Disposal: Two Social Criteria," *Science* 195, 4273 (7 January 1977): 23-31; Luther J. Carter, "Radioactive Wastes: Some Urgent Unfinished Business," *Science* 195 (18 February 1977):661-66 and 704; G. de Marsily et al., "Nuclear Waste Disposal: Can the Geologist Guarantee Isolation?" *Science* 197, 4303 (5 August 1977):519-27; P.A. Witherspoon et al., "Geologic Storage of Radioactive Waste: Field Studies in Sweden," *Science* 211 (27 February 1981):894-900; Isaac J. Winograd, "Radio-active Waste Disposal in Thick Unsaturated Zones," *Science* 212, 4502 (26 June 1981): 1457-64; John D. Bredehoeft and Tidu Maini, "Strategy for Radio-Active Waste Disposal in Crystalline Rocks," *Science* 213, 4505 (17 July 1981): 293-96; *American Scientist* 70 (March-April 1982):180-207.

2. For SSD literature, see the following: W.P. Bishop and Charles D. Hollister, "Seabed Disposal—Where to Look," *Nuclear Technology* 24 (December 1974):425-43; "High-Level Nuclear Wastes in the Seabed?" *Oceanus* 20, 1 (Winter 1977); Paul Grunwood and Geoffrey Webb, "Can Nuclear Wastes Be Buried at Sea?" *New Scientist,* 24 March 1977, pp. 709-11; David Deese, *Nuclear Power and Radioactive Waste: A Sub-Seabed Disposal Option?* (Cambridge, MA: Lexington Books, 1978); Robert R. Hessler and Peter A. Jumars, "The Relation of Benthic Communities to Radioactive Waste Disposal in the Deep Sea," *Ambio Special Report,* no. 6 (1979):93-96; Thomas C. Jackson, ed., *Nuclear Waste Management: The Ocean Alternative* (New York: Pergamon Press, 1981); Charles D. Hollister et al., "Subseabed Disposal of Nuclear Wastes," *Science* 213, 4514 (18 September 1981): 1321-26; K.R. Hinga, et al., "Disposal of High-Level Radioactive Wastes by Burial in the Sea Floor," *Environmental Science and Technology* 16, 1 (January 1982):28A-37A; Sandia National Laboratories, *Annual Reports of the Subseabed Disposal Program* (since 1976), and annual reports of the *Proceedings of the Seabed Working Group,* Nuclear Energy Agency (since 1976).

3. The most up-to-date assessment of the U.S. program so far is contained in Sandia National Laboratories, Seabed Programs Division 6334, *The Subseabed Disposal Program: 1983 Status Report,* Report No. SAND 83-1387 (October 1983).

Chapter 2: Radioactive Waste: Historical and Technical Dimensions

1. Kai N. Lee, "Radioactive Waste," in *The Nuclear Almanac*, ed. Jack B. Dennis (Reading, MA: Addison-Wesley, 1984).

2. U.S. Office of Technology Assessment, *Managing Commercial High-Level Radioactive Waste (Summary)*; OTA-O-172 (April 1982), p. 21.

3. Committee on Radioactive Waste Management, Panel on Hanford Wastes, *Radioactive Wastes at the Hanford Reservation, a Technical Review* (Washington, D.C.: National Academy of Sciences, 1978), p. 9.

4. Harmut Krugmann and Frank von Hippel, "Radioactive Wastes: A Comparison of U.S. Military and Civilian Inventories," *Science* 187 (1977):883-85.

5. Committee on Radioactive Waste Management, *Radioactive Wastes*, pp. 30, 36.

6. U.S. DOE, Assistant Secretary for Nuclear Energy, Nuclear Waste Management Program, *Spent Fuel and Radioactive Waste Inventories and Projections as of December 31, 1980*; DOE/NE-0017 (September 1981), p. 21.

7. Ronnie D. Lipschutz, *Radioactive Waste: Politics, Technology and Risk* (Cambridge, MA: Ballinger, 1980), pp. 132-34.

8. U.S. DOE, *Spent Fuel*, p. 62.

9. U.S. DOE, *Management of Commercially Generated Radioactive Waste;* DOE/EIS-0046 (1980). For an authoritative account of the state of scientific knowledge on the development of geologic repositories, see the NAS WISP Study.

10. NAS WISP Study.

11. Cyrus Klingsberg and James Duguid, *Status of Technology for Isolating High-Level Radioactive Wastes in Geologic Repositories;* DOE/TIC 11207 (draft); U.S. DOE, Office of Nuclear Waste Isolation (October 1980); Luther J. Carter, "Nuclear Wastes: The Science of Geologic Disposal Seen as Weak," *Science* 200 (1980):166-70; and de Marsily et al.

12. Witherspoon et al.

13. The NAS WISP Study argues that under existing Nuclear Regulatory Commission (NRC) regulations the geologic medium is still the primary barrier (see pp. 234-41).

14. Klingsberg and Duguid, p. 6.

15. A.E. Ringwood, S.E. Kesson, N.G. Ware, W. Hibberson, and A. Major, "Immobilisation of High Level Nuclear Reactor Wastes in SNYROC," *Nature* 278 (15 March 1979).

16. See Committee on Radioactive Waste Management, Subcommittee for Review of the *KBS-11 Plan for Disposal of Spent Nuclear Fuel* (Washington, D.C.: National Academy of Sciences, 1980).

17. Bredehoeft and Maini.

18. Comptroller General of the United States, General Accounting Office, *The Nation's Nuclear Waste—Proposals for Organization and Siting;* EMD-79-77 (21 June 1979).

19. NAS WISP Study, p. 8.

20. NWPA, Subtitle A.

21. *Ibid.,* Sec. 136.

22. *Ibid.,* Sec. 301.

23. *Ibid.,* Secs. 116-18.

24. *Ibid.,* Sec. 115.

25. Dorothy Nelkin and Michael Pollak, *The Atom Besieged* (Cambridge, MA: MIT Press, 1981).

Chapter 3: Sub-Seabed Disposal: Technical Considerations and Uncertainties

1. Mark B. Triplett and Kenneth A. Solomon (with Charles B. Bishop and Robert C. Tyce), *Monitoring Technologies for Ocean Disposal of Radioactive Waste;* R-2773-NOAA (Santa Monica: Rand Corporation, January 1982), pp. 10-11.

2. *Ibid.,* p. 14.

3. *Ibid.,* p. 11, quoting A. B. Joseph et al., "Sources of Radioactivity and Their Characteristics," in *Radioactivity in the Marine Environment* (Washington: National Academy of Sciences, 1971).

4. Triplett and Solomon, p. 11.

5. *Ibid.,* pp. v, 15.

6. U.S. Department of the Navy, *Disposal of Decommissoned, Defueled Naval Submarine Reactor Plants;* draft environmental impact statement, 22 December 1982; announced at Federal Register 47, 57085 (1982).

7. W. Jackson Davis (with contributions from Jon Van Dyke, Daniel Hirsch, Mary Anne Magnier, and Sherry P. Broder), "Evaluation of Oceanic Radioactive Dumping Programs"; unpublished draft (July 1982).

8. Jon Van Dyke, Kirk R. Smith, and Suliana Siwatibau, "Nuclear Activities and the Pacific Islanders"; unpublished draft (Honolulu: East-West Center, May 1983).

9. Center for Law and Social Policy and the Oceanic Society, "Joint Comments of Environmental and Other Citizen Organizations in Response to the Department of Navy's Draft Environmental Impact Statement on the Disposal of Decommissioned, Defueled Naval Submarine Reactor Plants" (Washington, 30 June 1983).

10. Estimated from Irvin C. Bupp, "The Actual Growth and Probable Future of the Worldwide Nuclear Industry," *International Organization* 35, 1 (Winter 1981): table 1, p. 60.

11. Sandia National Laboratories, *Subseabed Disposal Program Plan, Volume II: FY 1983 Budget and Subtask Work Plans;* SAND-83-0007/II (May 1983).

12. Our account of the origins of SSD is based upon an informal talk by Hollister at the Belmont Conference Center, Elkridge, Maryland, 25 July 1983, during a review of the SDP.

13. Bishop and Hollister, "Seabed Disposal."

14. See the discussion in NAS WISP Study, pp. 131-39.

15. See Davis.

16. See R. D. Klett, "Initial Phase of the Subseabed Thermal Sensitivity Studies"; in Sandia, *Annual Report . . . 1980,* Ap. B.

17. For an example of the current state of laboratory duplications, see B. T. Kenna, "Temperature and pH Effects on Sorption Properties of Subseabed Clay," in Sandia, *Annual Report . . . 1980,* Ap. F. Kenna's experiments were done at atmospheric pressure.

18. L. O. Olson and T. E. Ewart, "In-Situ Heat Transfer Experiment," in Sandia, *Annual Report . . . 1980,* Ap. F.

19. See Ap. AA, BB, CC, DD, and EE in Sandia, *Annual Report . . . 1980.*

20. *Ibid.,* pp. 42-43.

21. U. S. Office of Technology Assessment, p. 25.

22. Todd R. LaPorte, "Managing Nuclear Waste," *Society* 18 (July-August 1981): 60.

23. D. M. Talbert, "Subseabed Radioactive Waste Disposal Feasibility Program: Ocean Engineering Challenges for the '80s"; in *Oceans '80* (New York: Institute of Electrical and Electronic Engineers, 1980), p. 477.

24. Sandia, *Annual Report . . . 1980,* pp. 35-37.

Chapter 4: Potential Organizational and Operational Problems of the SSD Option

1. Talbert, "Subseabed Radioactive Waste Disposal Feasibility Program," p. 479.

2. We are indebted to James K. Asselstine for a helpful conversation on these points (Belmont Conference Center, Elkridge, Maryland, 27 July 1983).

3. Daniel M. Talbert, *Subseabed Disposal Program Annual Report, January to December 1979,* vol. 1, *Summary and Status;* Sandia Report No. SAND80-2577/1, p. 21.

4. James D. McClure, *The Probability of Spent Fuel Transportation Accidents;* Sandia Report No. SAND80-1971 (July 1981).

5. "Report of the System Analysis Task Group," in *Proceedings of the Sixth Annual NEA-Seabed Working Group Meeting* (Paris, 2-5 February 1981), ed. D. Richard Anderson; Sandia Report No. SAND81-0427, p. 38.

6. Talbert, "Subseabed Radioactive Waste Disposal Feasibility Program," p. 479.

7. *Ibid.*

8. James D. Thompson, *Organizations in Action* (New York: McGraw-Hill, 1967), pp. 14-18.

9. *Ibid.*, pp. 54-55.

10. *Ibid.*

11. LaPorte, "Managing Nuclear Waste," p. 60.

12. *Ibid.*

13. *Ibid.*, p. 61.

14. *Ibid.*

15. Todd R. LaPorte, "Nuclear Waste: Increasing Scale and Sociopolitical Impacts," *Science* 201 (7 July 1978):26-27.

16. Battelle Human Affairs Research Center, Pacific Northwest Division, *Nontechnical Issues in Waste Management: Ethical, Institutional and Political Concerns;* Report No. PNL-2400/UC-70 (May 1978), chs. 4 and 5.

17. *Ibid.*, pp. 44-45.

18. The notion of complexity is characterized by LaPorte and associates as "organized social complexity" in *Organized Social Complexity*, ed. Todd R. LaPorte (Princeton: Princeton University Press, 1975), p. 6.

19. Battelle describes control and management, monitoring, and information transfer (pp. 47-48). We have added operations.

20. Talbert, "Subseabed Radioactive Waste Feasibility Program," p. 479.

21. Chris C. Demchak, "Complexification and Organizational-Institutional Design: The Case of the U.S. Radioactive Waste Program"; unpublished report no. IGS/RW-004, Institute of Governmental Studies, University of California, Berkeley, November 1980.

22. Todd R. LaPorte, personal communication, 7 June 1982.

Chapter 5: Potential Domestic Political Problems Connected with the SSD Option

1. Information provided in this chapter, unless otherwise noted, is based on a series of nonattributable interviews of policy-level representatives in government, industry, the environmental and public-interest community, and the scientific community.

2. A major research effort in this area is J. A. Hebert et al., *Nontechnical Issues in Waste Management: Ethical, Institutional, and Political Concerns;* Battelle

Human Affairs Research Center, Report No. PNL-2400 (1978). The authors have identified the following waste management issues: dislocation of costs and/or benefits to future generations, need for candor, public involvement, uncertainty, risk and equity issues, safeguards and civil liberties, conservation and alternative power production, transportation, institutional issues, irreversible decisions, commercial versus defense wastes, international responsibilities regarding waste management and reprocessing, and costs of waste management.

3. Ted Peters, "Ethical Considerations Surrounding Nuclear Waste Repository Siting and Mitigation"; unpublished manuscript (Berkeley: Pacific Lutheran Theological Seminary and Graduate Theological Union, 1981). See also Ted Peters, "The Ethics of Radwaste Disposal," *Christian Century* 99 (1982):271-73.

4. Peters, "Ethical Considerations."

5. U.S. Congress, House, Committee on Interior and Insular Affairs, *Radiological Contamination of the Oceans;* oversight hearings before the Subcommittee on Energy and the Environment, 94th Cong., 2d Sess. (1976); and U.S. Congress, House, Committee on Merchant Marine and Fisheries, *Hearings on Nuclear Waste Disposal* before the Subcommittee on Oceanography, 96th Cong., 2d Sess. (1980).

6. 33 U.S.C. 1402(f) and 1412(a).

7. The convention was ratified 29 December 1972 and in force 30 August 1975; in ILM 11 (1972):1291, arts. 3 and 4.

8. Statement of Louis Harris before the Sub-Committee on Health and the Environment, House Committee on Energy and Commerce, 15 October 1981.

9. California Senate Joint Resolution No. 27, introduced by Senators Barry Keene and Milton Marks, 10 September 1981. The resolution was passed unanimously by the Senate Rules Committee on 13 January 1982. Among the reasons for requesting the ban on waste disposal in the Pacific, the following are listed in the resolution: past radioactive waste dumping (of LLW), the Navy's plans to scuttle decommissioned submarines, and DOE development of the SSD option.

10. "Matsunaga Calls for Firm U.S. Policy against Nuclear Testing and Dumping in the Pacific"; Press Release from U.S. Senator Matsunaga, 18 February 1982.

11. Widely cited to support this contention is the U.S. General Accounting Office Report, *Hazards of Past Low-Level Radioactive Waste Ocean Dumping Have Been Overemphasized* (Gaithersburg, MD: U.S. General Accounting Office, 1981).

12. Colin Norman, "U.S. Considers Ocean Dumping of Radwastes," *Science* 215 (1982):1217-19.

13. *Ibid.*

14. For further discussion of the importance of assigning burden of proof in environmental litigation, see William H. Rodgers, *Handbook on Environmental Law* (St. Paul, MN: West Publishing, 1977).

15. Hebert et al., *Nontechnical Issues in Waste Management*, p. 14.

16. R.H. Johnson, Jr., "Public Understanding of Radiation: What's the Problem," *Proceedings of the Public Meetings 10/11/80 "To Address a Proposed Federal Research Agenda,"* vol. 1, *Issue Papers* (Interagency Radiation Research Committee).

17. Peters, *Ethical Considerations*, p. 59.

18. *Ibid.*

19. See, for example, "Our Nuclear Policies in the Pacific," *Honolulu Star-Bulletin*, 20 February 1982.

20. See, for example, Statement of Clifton E. Curtis before the Sub-Committee on Oceanography, House Committee on Merchant Marine and Fisheries, 20 November 1980. The statement was endorsed by twenty-two environmental and public-interest groups.

21. For details on this resolution, see note 9 above.

22. Deese, p. 14.

Chapter 6: Potential International Problems Connected with the SSD Option

1. Hebert et al., pp. 13-14.

2. Rochlin, *Plutonium, Power and Politics*, p. 328.

3. Doc. A/CONF. 62/122, 7 October 1982, esp. Part VI dealing with the Continental Shelf.

4. General Assembly Resolution 2749 (25), 17 December 1970.

5. Martine Rémond-Gouilloud, "Pollution from Seabed Activities," in *The Environmental Law of the Sea*, ed. Douglas M. Johnston (Glaud, Switzerland: IUCN, 1981), pp. 254-55.

6. Deese; David A. Deese et al., *Political and Institutional Implications of the Seabed Assessment Program for Radioactive Waste Disposal;* Final Report for the Seabed Assessment Program, Sandia National Laboratories; prepared by Urban Systems Research and Engineering, 1981.

7. See the following (inter alia): statement by Elliott Richardson (former ambassador to Great Britain), "Sub-Seabed Disposal in the Context of the Law of the Sea," in *Nuclear Waste Management: The Ocean Alternative,* ed. Thomas C. Jackson (New York: Pergamon Press, 1981), pp. 85-86; Jean-Pierre Queneudec, "The Effects of Changes in the Law of the Sea on Legal Regimes Relating to the Disposal of Radioactive Waste in the Sea"; unpublished manuscript prepared for the NEA, 28 January 1982; and John

Norton Moore, "Some Preliminary Considerations Concerning the Legal and Foreign Policy Aspects of a Regime for Subseabed Disposal of Nuclear Wastes"; memorandum prepared for Sandia National Laboratories, 22 January 1982.

8. In the public record this position was articulated by Richardson.

9. Moore.

10. Kenneth R. Hinga, "The Conflicts Between Deep Ocean Mining and Subseabed Disposal of Radioactive Wastes"; unpublished manuscript, Sandia National Laboratories, Report No. SAND81-1486J.

11. For assessments of current trends in the anti-nuclear movement, see the following: Robert Cameron Mitchell, "From Elite Quarrel to Mass Movement," *Society*, July/August 1981:76-84; Dorothy Nelkin and Michael Pollak, "Nuclear Protest and National Policy," *Society*, July/August 1981: 34-38; Geoffrey Murray, "Japan's Rising Peace Movement: Will Antinuclear Campaign Weaken U.S. Tie?" *Christian Science Monitor*, 6 April 1982, p. 1; U.S. Senate, Committee on Energy and Natural Resources, *Hearing on Pacific Spent Nuclear Fuel Storage*, 96th Cong., 1st Sess., 5 June 1979; Robert Trumbull, "Pacific Governors Oppose Dumping Atom Wastes," *New York Times*, 5 October 1980; and Angela Gennino, "Nuclear Protests Across the Pacific," *Not Man Apart*, June 1981:10-11 and 15.

12. Our discussion of regimes draws on the following works: Robert O. Keohane and Joseph S. Nye, *Power and Interdependence* (Boston: Little, Brown, 1977), pp. 8-11, 35-37, 55-57, 124-27, 146, 148, and 151; John Gerard Ruggie, "International Responses to Technology: Concepts and Trends," *International Organization* 29, 3 (Summer 1975):557-84; Oran Young, "International Regimes: Problems of Concept Formation," *World Politics* 32 (1980):331-56; Ernst Haas, "Why Collaborate?" Issue-Linkage and International Regimes," *World Politics* 32 (1980):375-405.

13. Haas, p. 397.

14. *Ibid.*, pp. 398-99.

15. Robert O. Keohane and Joseph S. Nye, "Transgovernmental Relations and International Organizations," *World Politics* 27, 1 (October 1974):39-62.

16. Robert O. Keohane and Joseph S. Nye, eds., *Transnational Relations and World Politics;* special issue of *International Organization* 25, 3 (Summer 1971):329-50.

17. Keohane and Nye, *Power and Interdependence*, pp. 35-37.

18. *Ibid.*, p. 146.

19. *Ibid.*, pp. 55-57.

20. Haas, p. 371.

21. *Ibid.*

22. *Ibid.*

23. *Ibid.*, p. 372.

24. *Ibid.*, p. 374.

25. See General Assembly Resolution 1721 (XVI), "International Co-operation in the Peaceful Uses of Outer Space," 20 December 1961, and General Assembly Resolution 1802 (XVII), "International Co-operation in the Peaceful Uses of Outer Space," 14 December 1962.

26. Edward Miles, "Transnationalism in Space: Inner and Outer," in Keohane and Nye, eds., p. 603.

27. See J.D. Kingham and D.M. McRae, "Competent International Organizations and the Law of the Sea," *Marine Policy* (April 1979): 106-32, and Edward Miles, "On the Roles of International Organizations in the New Ocean Regime," in *The Law of the Sea in the 1980's*, ed. Choon-ho Park (Honolulu: University of Hawaii Press, 1983), pp. 383-445.

28. See Rochlin, *Plutonium, Power and Politics,* pp. 194, 197, 226-27, and 303-4.

29. See Miles, "On the Roles of International Organizations in the New Ocean Regime."

Epilogue

1. See OECD/NEA, "Legal Aspects of the Disposal of Radioactive Waste under the Seabed"; paper presented to the Ad Hoc Group of Legal Experts on Dumping, International Maritime Organization, 12-14 December 1983; DOC. LDC/LG.2/3/1. See also the summary of comments made by the OECD/NEA observers in *The Dumping of Radioactive Wastes at Sea: Report of the Ad Hoc Group of Legal Experts on Dumping,* DOC. LDC/8/5/3 (19 December 1983), pp. 5-6.

2. Resolution LDC 1/7/4/Rev. 1 (17 February 1983).

3. Resolution LDC 7/WP-9 (17 February 1983).

4. See *The Dumping of Radioactive Wastes at Sea,* p. 16.

5. *Draft Report of the Eighth Consultative Meeting of the Contracting Parties to the Convention on the Prevention of Marine Pollution by Dumping of Wastes and Other Matter,* IMO; DOC. LDC8/WP.9/Add. 2 (23 February 1984), p. 9.

INDEX

Acceptability of SSD, 34, 36, 42, 49, 55, 60-61, 64-67, 94-95
Accidents: effects of, 27, 67, 73-74, 76; risks of, 34, 40, 48, 65-66; safeguards against, 66; scenarios for, 40-41
Ad Hoc Group of Legal Experts on Dumping, 94-95
AEC. *See* Atomic Energy Commission
Anderson, D. Richard, 23, 35n
Antarctic resources regimes, 73-74
Anti-dumping coalition, 94-94. *See also* Opposition to SSD
Anti-nuclear movement, 6, 29, 52, 56, 64, 74-75, 77-78, 80-81, 83-84. *See also* Opposition to SSD
Atmospheric testing of nuclear weapons, 20
Atomic Energy Commission (AEC), 12-13
Away From Reactor (AFR) storage facilities, 39

Battelle Human Affairs Research Center, 43, 67
Belgium, 82-83
Bishop, William P., 21, 23
Borosilicate, 39
Bowen, Vaughn T., 21, 23
Bredehoeft, J.D., 16
Breeder reactor, 51

California, 57, 62
Canisters, 24-26, 67
Carter administration, 78
China, People's Republic of, 69, 71
Coalition-building, 4, 50, 56-57, 62-63, 85, 91, 93. *See also* Political actors in SSD

Coastal state jurisdiction, 68, 76-77
Coast Guard (U.S.), 39
Commerce, Department of, 57
Commercial interest groups. *See* Nuclear industry lobby
Commission of the European Communities, 21
"Common heritage" principle, 5, 56, 61-62, 64, 68, 73-74
Compensation. *See* Liability of SSD
Computer simulation models of SSD. *See under* Sub-seabed disposal
Congress (U.S.), 4, 51-52, 58
Continental shelf limits, 73
Convention on Liability for Damage Caused by Objects Launched into Outer Space, 69
Convention on the Prevention of Marine Pollution by Dumping of Wastes and Other Matter (1972). *See* London Convention
Cost-effectiveness, 56, 61

Deep drilling. *See* Emplacement techniques
Deese, David, 63
Department of State (U.S.), 52-53, 57-58
Disposal alternatives, 1, 2, 14-16
DOE. *See* Energy, Department of (U.S.)
"Dumping at sea." *See* London Convention

Education for SSD, 72, 81, 87, 89
EEZ. *See* Exclusive Economic Zone
Emplacement techniques, 27, 40-41, 82-84, 88. *See also* Retrievability

EDWARD MILES is Professor of Marine Studies and Public Affairs and Director of the Institute for Marine Studies, University of Washington. Trained as a political scientist, he now works primarily on policy problems of ocean use, particularly those concerning fisheries management, marine scientific research, radioactive waste disposal in the ocean, and the law of the sea. He is a participant in the Seabed Disposal Program, managed for the U.S. Department of Energy by Sandia National Laboratories, Inc. and the Seabed Working Group, Nuclear Energy Agency, OECD.

KAI LEE is Associate Professor of Environmental Studies and Political Science, University of Washington. Trained in both experimental physics and political science, he works on various problems relating to alternative forms of energy production and use. He is a member of the Board on Radioactive Waste Management of the National Research Council.

ELAINE CARLIN is Program Manager for the Low-Level Radioactive Waste Program of the Department of Ecology, State of Washington. Trained in both the marine sciences and marine policy, she pursued graduate study at the Institute for Marine Studies, University of Washington.

INSTITUTE OF INTERNATIONAL STUDIES
UNIVERSITY OF CALIFORNIA, BERKELEY

215 Moses Hall   Berkeley, California 94720

CARL G. ROSBERG, *Director*

Monographs published by the Institute include:

RESEARCH SERIES

1. *The Chinese Anarchist Movement.* R.A. Scalapino and G.T. Yu. ($1.00)
7. *Birth Rates in Latin America.* O. Andrew Collver. ($2.50)
15. *Central American Economic Integration.* Stuart I. Fagan. ($2.00)
16. *The International Imperatives of Technology.* Eugene B. Skolnikoff. ($2.95)
17. *Autonomy or Dependence in Regional Integration.* P.C. Schmitter. ($1.75)
19. *Entry of New Competitors in Yugoslav Market Socialism.* S.R. Sacks. ($2.50)
20. *Political Integration in French-Speaking Africa.* Abdul A. Jalloh. ($3.50)
21. *The Desert & the Sown: Nomads in Wider Society.* Ed. C. Nelson. ($5.50)
22. *U.S.-Japanese Competition in International Markets.* J.E. Roemer. ($3.95)
23. *Political Disaffection Among British University Students.* J. Citrin and D.J. Elkins. ($2.00)
24. *Urban Inequality and Housing Policy in Tanzania.* Richard E. Stren. ($2.95)
25. *The Obsolescence of Regional Integration Theory.* Ernst B. Haas. ($4.95)
26. *The Voluntary Service Agency in Israel.* Ralph M. Kramer. ($2.00)
27. *The SOCSIM Microsimulation Program.* E. A. Hammel et al. ($4.50)
28. *Authoritarian Politics in Communist Europe.* Ed. Andrew C. Janos. ($3.95)
29. *The Anglo-Icelandic Cod War of 1972-1973.* Jeffrey A. Hart. ($2.00)
30. *Plural Societies and New States.* Robert Jackson. ($2.00)
31. *Politics of Oil Pricing in the Middle East, 1970-75.* R.C. Weisberg. ($4.95)
32. *Agricultural Policy and Performance in Zambia.* Doris J. Dodge. ($4.95)
33. *Five Classy Computer Programs.* E.A. Hammel & R.Z. Deuel. ($3.75)
34. *Housing the Urban Poor in Africa.* Richard E. Stren. ($5.95)
35. *The Russian New Right: Right-Wing Ideologies in USSR.* A. Yanov. ($5.95)
36. *Social Change in Romania, 1860-1940.* Ed. Kenneth Jowitt. ($4.50)
37. *The Leninist Response to National Dependency.* Kenneth Jowitt. ($4.95)
38. *Socialism in Sub-Saharan Africa.* Eds. C. Rosberg & T. Callaghy. ($12.95)
39. *Tanzania's Ujamaa Villages: Rural Development Strategy.* D. McHenry. ($5.95)
40. *Who Gains from Deep Ocean Mining?* I.G. Bulkley. ($3.50)
41. *Industrialization & the Nation-State in Peru.* Frits Wils. ($5.95)
42. *Ideology, Public Opinion, & Welfare Policy: Taxes and Spending in Indus- dustrialized Societies.* R.M. Coughlin. ($6.50)
43. *The Apartheid Regime: Political Power and Racial Domination.* Eds. R.M. Price and C. G. Rosberg. ($12.50)
44. *Yugoslav Economic System in the 1970s.* L.D. Tyson. ($5.50)
45. *Conflict in Chad.* Virginia Thompson & Richard Adloff. ($7.50)
46. *Conflict and Coexistence in Belgium.* Ed. Arend Lijphart. ($7.50)